Winfried Hartmeier

Immobilized
Biocatalysts

An Introduction

Translated by Joy Wieser

With 115 Figures

Springer-Verlag
Berlin Heidelberg New York
London Paris Tokyo

Prof. Dr. Winfried Hartmeier
Department of Technical Biochemistry
University of Hohenheim
7000 Stuttgart 70, FRG

Translator:
Joy Wieser
Madleinweg 19
6064 Rum/Innsbruck, Austria

Original German edition: W. Hartmeier, Immobilisierte Biokatalysatoren
© Springer-Verlag Berlin Heidelberg 1986

Legend of cover motif: Propeller-Loop reactor with immobilized biocatalysts

---

ISBN 3-540-18808-8 Springer-Verlag Berlin Heidelberg New York
ISBN 0-387-18808-8 Springer-Verlag New York Berlin Heidelberg

Library of Congress Cataloging-in-Publication Data. Hartmeier, Winfried. Immobilized biocatalysts. Translation of Immobilisierte Biokatalysatoren. Includes index. 1. Immobilized enzymes–Industrial applications 2. Immobilized enzymes–Synthesis. I. Title. TP248.E5H3813 1988 661'.8 88-489

© by Springer-Verlag Berlin Heidelberg 1988
Printed in Germany

Offsetprinting and bookbinding: Druckhaus Beltz, 6944 Hemsbach/Bergstr.
2132/3130-543210

Author's Preface to the English Edition

The appearance of an English translation one year after the publication of the first German edition is due to the conviction of publisher and author that, despite the wealth of Anglo-American literature concerning immobilized catalysts, there is a lack of inexpensive introductory books on the subject.

Some slight extensions have been made to the original German text. In the practical section, two additional experiments introduce the student to recent techniques for using membrane reactors and for biocatalyst encapsulation using liquid membranes. Some of the most important publications that have appeared in the meantime have been taken into consideration and added to the list of recommended literature.

I am greatly indebted to the translator, Mrs. Joy Wieser, and to Springer-Verlag, especially to Dr. Dieter Czeschlik, who encouraged and supported me in many ways.

W. Hartmeier

Contents

THEORETICAL SECTION

1 General Principles

1.1 Principles of Biocatalysis

Enzymes are the biocatalytically active entities upon which the metabolism of all living organisms is based. They speed up (bio)chemical reactions by lowering the energy of activation, without themselves appearing in the reaction products. In this, and in the fact that the catalyst itself is not used up, the action of enzymes resembles that of inorganic catalysts.

For the continued existence of organic compounds, and hence for the presence of life on Earth, the existence of an activation barrier is indispensable for preventing continual breakdown. At moderate temperatures many substances are metastable; in other words, they do not break down even though their energy content is considerably higher than that of their breakdown products. Not until an adequate stimulus is provided by the addition of energy, or the energy of activation is sufficiently lowered by a catalyst, are such substances transformed at a greater speed. Figure 1 and Table 1 illustrate the effect of enzymes and inorganic catalysts on the energy of activation and the speed of reaction. The breakdown of hydrogen peroxide into water and oxygen has been taken as an example.

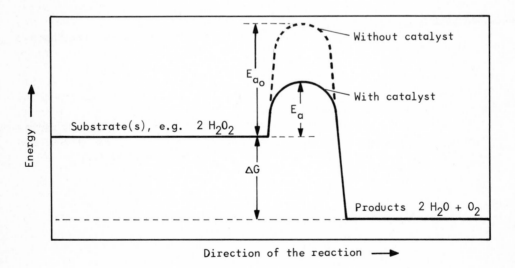

Fig. 1. Energy in a reaction with and without catalyst

Table 1. Breakdown of hydrogen peroxide with and without catalyst

Catalyst	Energy of activation kJ/mol	Relative reaction rate
Without catalyst	75.4	1
Platinum (inorganic)	50.2	$2 \cdot 10^4$
Catalase (enzyme)	8.4	$3 \cdot 10^{11}$

The catalytic action of enzymes and inorganic catalysts involves their ability to alter the distribution of charges on the compound to be converted, thus bringing about a lowering of the energy of activation, E_a (cf. Fig. 1). As a rule, enzymes are more efficient in this respect, i.e., they lead to a much greater decrease in the energy of activation than inorganic catalysts (cf. Table 1). This is why enzyme-catalyzed reactions usually proceed under mild conditions, i.e. at lower temperatures, atmospheric pressure and physiological pH values. Furthermore, in contrast to inorganic catalysts, they are highly specific; in other words, a particular enzyme usually catalyzes only one reaction. This means that, to a large extent, side reactions can be avoided by employing enzymatic breakdown.

The thermodynamic equilibrium of a reaction is in no way affected by the use of catalysts, whether inorganic or enzymic. In the presence of a catalyst, the state of equilibrium is simply reached sooner. In nature, however, for instance in the living cell, it is quite possible for reactions to take place even if the reaction products have considerably higher free energy (G) than the initial substrates. This is achieved by the coupling of such an energy-consuming (endergonic) reaction, whose ΔG value is positive, with an energy-liberating (exergonic) reaction, where the second reaction must be sufficiently exergonic (negative ΔG) for the sum of the changes in free energy of the two reactions, ΔG, to be zero or negative.

From what has been said so far it is clear that enzymes are specific biocatalysts. The term biocatalyst applies not only to single enzymes, but also includes chains of enzymes linked to form larger units. Even a cell, with its vast number of different enzymes, can be regarded as a complex biocatalyst capable, for example, of transforming sugar into ethanol and carbon dioxide, or of even more complicated biosynthetic feats.

1.2 Structure of Enzymes

All enzymes, as far as we know, are proteins, although by no means every protein is an enzyme. Examples of catalytically inactive proteins are the antibodies, whose role is protective, or structural proteins which serve for support.

In order to develop catalytic properties some enzyme proteins require the cooperation of a low-molecular nonprotein substance or group. Such nonprotein accessory substances are called coenzymes or cofactors, unless they are tightly bound to the enzyme protein, in which case they are often termed prosthetic groups. The protein portion or moiety is known as the apoenzyme and the combination of enzyme protein and nonprotein active group is the holoenzyme.

Not always does the nonprotein moiety of an enzyme play a direct role in biocatalysis. Many enzymes, especially those of technical importance, are glycoproteins, which means that they possess a carbohydrate moiety which, as a rule, is not involved in the catalytic action. Yeast invertase is an enzyme of this kind: about half of its molecular weight is due to the polysaccharide mannan.

Protein Moiety

The protein moiety of an enzyme consists of amino acids which, with the exception of proline and hydroxyproline (see Table 2), have the following general formula:

$$H_2N-\underset{\underset{R}{|}}{CH}-COOH$$

The splitting off of water between the α-carboxyl group of one amino acid and the α-amino group of another gives rise to a dipeptide.

$$H_2N-\underset{\underset{R_1}{|}}{CH}-COOH \;+\; H_2N-\underset{\underset{R_2}{|}}{CH}-COOH \xrightarrow{\qquad\qquad} H_2N-\underset{\underset{R_1}{|}}{CH}-CO-NH-\underset{\underset{R_2}{|}}{CH}-COOH$$
$$H_2O$$

The linkage between the amino acids is termed an amide link or a peptide bond, and the amino acids thus linked are known as residues. By means of further condensation, i.e., formation of a chain accompanied by the elimination of water, large molecules known as polypetides are formed. It seems that a chain length of about 50 amino acids is necessary for enzyme activity, although the number of residues is usually much higher. Up to several thousands of amino acid units are often found in enzyme proteins, giving a molecular weight of 5000 to several million. Table 2 shows the 20 amino acids generally involved in the structure of enzymes.

Table 2. The most important amino acids

$COOH$ $\mid$ H_2N-C-H $\mid$ H Glycine (Gly)	$COOH$ $\mid$ H_2N-C-H $\mid$ CH_3 L-Alanine (Ala)	$COOH$ $\mid$ H_2N-C-H $\mid$ CH $H_3C\;\;CH_3$ L-Valine (Val)	$COOH$ $\mid$ H_2N-C-H $\mid$ CH_2 $\mid$ CH $H_3C\;\;CH_3$ L-Leucine (Leu)	$COOH$ $\mid$ H_2N-C-H $\mid$ CH $H_2C\;\;CH_3$ H_3C L-Isoleucine (Ile)
$COOH$ $\mid$ H_2N-C-H $\mid$ CH_2 L-Phenylalanine (Phe)	H_2C-CH_2 $H_2C\;\;\;\;CH-COOH$ NH L-Proline (Pro)	$COOH$ $\mid$ H_2N-C-H $\mid$ CH_2 $\mid$ OH L-Serine (Ser)	$COOH$ $\mid$ H_2N-C-H $\mid$ $H-C-OH$ $\mid$ CH_3 L-Threonine (Thr)	$COOH$ $\mid$ H_2N-C-H $\mid$ CH_2 $\mid$ SH L-Cysteine (Cys)
$COOH$ $\mid$ H_2N-C-H $\mid$ CH_2 $\mid$ CH_2 $\mid$ S $\mid$ CH_3 L-Methionine (Met)	$COOH$ $\mid$ H_2N-C-H $\mid$ CH_2 $\mid$ C $\parallel$ CH NH L-Tryptophane (Trp)	$COOH$ $\mid$ H_2N-C-H $\mid$ CH_2 OH L-Tyrosine (Tyr)	$COOH$ $\mid$ H_2N-C-H $\mid$ CH_2 $\mid$ $COOH$ L-Aspartic acid (Asp)	$COOH$ $\mid$ H_2N-C-H $\mid$ CH_2 $\mid$ CH_2 $\mid$ $COOH$ L-Glutamic acid (Glu)
$COOH$ $\mid$ H_2N-C-H $\mid$ CH_2 $\mid$ $CONH_2$ L-Asparagine (Asn)	$COOH$ $\mid$ H_2N-C-H $\mid$ CH_2 $\mid$ CH_2 $\mid$ $CONH_2$ L-Glutamine (Gln)	$COOH$ $\mid$ H_2N-C-H $\mid$ CH_2 $\mid$ CH_2 $\mid$ CH_2 $\mid$ CH_2-NH_2 L-Lysine (Lys)	$COOH$ $\mid$ H_2N-C-H $\mid$ CH_2 $\mid$ $CH_2\;\;NH_2$ $H_2C\;\;\;C=NH$ NH L-Arginine (Arg)	$COOH$ $\mid$ H_2N-C-H $\mid$ CH_2 $\mid$ $C-N$ $HC\;\;\;\;CH$ NH L-Histidine (His)

The sequence in which the amino acids of an enzyme protein are linked to one another is termed its primary structure. In describing the makeup of a polypeptide the convention is to begin with the N-terminal, i.e., the amino acid whose amino group is unbound, and to employ the abbreviations shown in brackets in Table 2. The amino acid sequence of many enzymes important in industry has already been elucidated. Figure 2 shows as an example the amino acid sequence of papain, a plant protease used in stabilizing beer (chill-proofing) and in tenderizing meat.

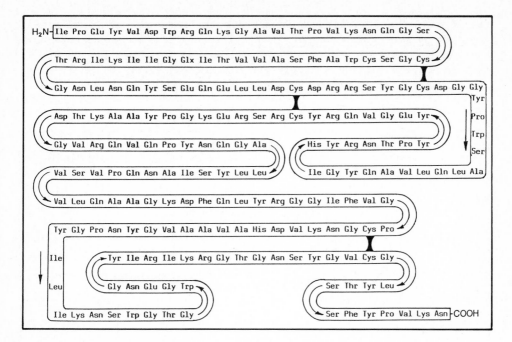

Fig. 2. Sequence of amino acids of papain

Conformation and Active Center

As flexible structures, polypeptide chains have a tendency to adopt
characteristic spatial conformations due to the steric properties and
charge distribution of their constituent amino acids, as well as to their
ability to form hydrogen bonds between the carbonyl-oxygen and amide-
nitrogen atoms. A common arrangement of this kind is the α-helix, a cork-
screw structure comprising 18 residues per 5 turns. Another arrangement,
less often found in enzymes, is the ß-structure or ß-sheet. α-helix and ß-
sheet structure are also known as the secondary structure of a protein.

In addition to α-helix and ß-sheet regions there are always some
seemingly random structural conformations in the polypeptide chain of an
enzyme molecule. In order for a protein to exert a specific enzymatic
action at least some regions in each polypetide chain must have a certain
typical spatial conformation. This confers upon the enzyme its own characteristic form and activity. The spatial structure assumed by an enzyme as
a result of interaction between the amino acid side chains is termed the
tertiary structure: it is stabilized partly by disulfide bridges that form
between the cysteine residues of near-lying portions of the chain (cf.
Fig. 2). To a very considerable extent the conformation assumed by an
enzyme protein and the stabilization of its structure is due to ionic
bonds, hydrophobic interaction, hydrogen bonds and van der Waals forces.

Some enzymes consist of several, often two or four, identical or
different polypeptide subchains spatially arranged to give a specific, so-

called quaternary structure. Such a complex structure is usually held together by noncovalent bonds, particularly by ionic bridges, hydrogen bonds or hydrophobic interactions. The quaternary structure is very important in enzymes playing a regulatory role in metabolism (allosteric enzymes), but is seldom present in enzymes of industrial importance.

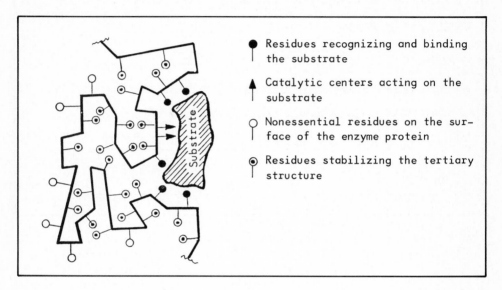

● Residues recognizing and binding the substrate

▲ Catalytic centers acting on the substrate

○ Nonessential residues on the surface of the enzyme protein

◉ Residues stabilizing the tertiary structure

Fig. 3. Scheme of an enzyme protein with its active site

In a certain region of every enzyme the amino acids are arranged in a specific sequence and spatial conformation that are responsible for the catalytic function. This region is known as the active site or catalytic center. It is here that the substrate is bound to the enzyme and broken down into the reaction products. Only a very few amino acid residues participate directly in the catalytic action although most of the other amino acid side chains are also important and their indirect effect is in many cases even essential for biocatalysis to take place. Some of them are involved in the recognition and binding of the substrate molecule to the active site, others participate in the formation and stabilization of the three-dimensional conformation of the enzyme molecule. Figure 3 shows a schematized section of an amino acid chain with its active site, and amino acid side chains with their various functions.

Coenzymes

Coenzymes or cofactors are low-molecular, nonprotein molecules whose presence is essential for many enzyme reactions. There are two types of coenzymes: those that dissociate from their apoenzymes after completion of the reaction and those that remain associated. In cases where a cofactor is relatively tightly bound to its apoenzyme it can be termed a prosthetic group. Unfortunately, it is not always easy to make a distinction between the two types of coenzyme since transitional degrees of binding and dissociation are also encountered. For the purposes of immobilization, however, permanence of association between co- and apoenzyme is of crucial importance because this determines whether or not special procedures have to be adopted for coenzyme immobilization (cf. Sect. 2.8). In Table 3 some examples of freely dissociating (above broken line) and tightly bound cofactors (below broken line) are listed.

Table 3. Examples of coenzymes

Abbreviation and full name		Function	Corresponding apoenzymes
NAD	Nicotinamide-adenine dinucleotide	Transfer of hydrogen	Dehydrogenases
NADP	Nicotinamide-adenine dinucleotide phosphate	Transfer of hydrogen	Dehydrogenases
ATP	Adenosine-triphosphate	Phosphorylations, transphosphorylations	Kinases, transferases, synthetases
CoA	Coenzyme A	Transfer of acyl residues	Acyl transferases, thiol ligases
FAD	Flavin-adenine dinucleotide	Transfer of hydrogen	Oxidases
PAL	Pyridoxal phosphate	Transamination, decarboxylation	Transaminases, decarboxylases
TPP	Thiamine pyrophosphate	Decarboxylation	Decarboxylases
---	Heme coenzymes	Electron transport	Monooxygenases, mutases peroxidases
---	Biotin	Transfer of CO_2	Carboxylases

Most of the coenzymes listed in Table 3 are synthesized from vitamins. For example NAD contains nicotinic acid and the major part of FAD consists of riboflavin (vitamin B_2).

Metals are often involved in the catalytic function as inorganic complements of enzyme reactions. These metals are not termed coenzymes, but rather cofactors. Thus, the term cofactor is used in a more general and wider sense than coenzyme.

1.3 Classification and Nomenclature of Enzymes

In the early days of enzyme research the name given to an enzyme was largely a matter of chance or a whim of its discoverer. Inevitably, the result was a bewildering variety of names, sometimes even for one and the same enzyme. In 1956 the International Union of Biochemistry (IUB) called into being an Enzyme Commission (EC), followed by a Nomenclature Commission (NC), whose recommendations on nomenclature and classification of enzymes are recognized and adhered to throughout the scientific world. A few of the criteria involved are outlined below; further details can always be found in the most recent edition of *Enzyme Nomenclature* (International Union of Biochemistry, 1984).

Systematic Names

The principle employed is that the systematic name of an enzyme should include the substrate, the type of reaction to be catalyzed and the suffix "-ase." Thus, an enzyme catalyzing a simple monosubstrate reaction is named according to the following scheme:

substrate - type of reaction catalyzed - ase.

If the enzyme acts with two substrates both substrate names are included, separated by a colon. The general form of the name is then:

substrate A : substrate B - type of reaction - ase.

Abbreviations commonly used in biochemistry, such as NAD or ADP, are also used in enzyme names. For substrates that usually occur in the ionized form the name of the corresponding salt is used (e.g. pyruvate or succinate). Within the various classes of enzymes the rules governing the systematic names are more narrowly defined with respect to the particular type of reaction. In redox reactions (first main class) for example, the enzyme substrates are electron donors and acceptors, which results in the following systematic names:

electron donor : electron acceptor - oxidoreductase.

This can be shown more explicitly for alcohol dehydrogenase, which catalyzes the oxidation of alcohol to aldehyde, in the course of which reaction hydrogen is transferred to NAD. Thus alcohol is the electron donor and NAD the acceptor, which means that the systematic name for alcohol dehydrogenase is alcohol:NAD$^+$ oxidoreductase.

In the case of transferases, group donor and group acceptor replace the electron donor and acceptor. In addition, the type of reaction catalyzed is more specifically defined by the inclusion of the group transferred. The name of the transferase is thus:

$$\boxed{\text{donor : acceptor - transferred group - transferase.}}$$

Hydrolases are named quite simply as follows: in accordance with the basic principles stated at the outset, an enzyme of this class ought to be called "substrate-hydrolase." In fact, it is not always easy to recognize the exact nature of the substrate. Furthermore, if the substrate contains several groups that can be split off by hydrolysis the group actually hydrolyzed should also be included in the name as follows:

$$\boxed{\text{substrate - group split off - hydrolase.}}$$

In this way it is possible to make a clear distinction between different hydrolases with similar actions, such as, for example, the familiar enzymes ß-amylase and glucoamylase (or amyloglucosidase). Both enzymes split α-1,4-D-glucans, i.e., D-glucose molecules joined by α-1,4-glucosidic linkages to form polymers. ß-Amylase splits off maltose umits whereas glucoamylase splits off glucose, which is clearly indicated by the systematic names α-1,4-D-glucan maltohydrolase for ß-amylase and α-1,4-D-glucan glucohydrolase for glucoamylase. Table 4 lists further examples of nomenclature.

Trivial Names

Although the systematic names are scientifically exact and unambiguous, in many cases they are unfortunately too long for everyday purposes, so that trivial names are still in use. The 1984 edition of *Enzyme Nomenclature* (International Union of Biochemistry, 1984) even includes recommendations made by the Nomenclature Commission of the IUB on this topic. The name recommended is placed immediately after the EC number, thus being promoted to a very obvious position, preceding the systematic name. In many cases the name recommended is the same as that most commonly encountered in the literature. In a few cases, e.g., glucan 1,4-α-glucosidase for glucoamylase, it is to be hoped that authors will in future be guided by the recommendations. Table 4 gives the possible names of some common enzymes.

Table 4. Examples of the nomenclature of enzymes

EC No	Name recommended	Systematic name	Other names
1.1.1.1	Alcohol dehydrogenase	Alcohol:NAD$^+$ oxidoreductase	Aldehyde reductase
1.1.3.4	Glucose oxidase	ß-D-glucose:O$_2$ oxidoreductase	Glucose oxyhydrase, notatin, nigerin
1.11.1.6	Catalase	H$_2$O$_2$:H$_2$O$_2$ oxidoreductase	
2.4.1.5	Dextran sucrase	Sucrose:1,6-α-D-glucan 6-α-D-glucosyltransferase	Sucrose 6-glucosyltransferase
2.4.1.10	Levan sucrase	Sucrose:2,6-ß-D-fructan 6-ß-D-fructosyltransferase	Sucrose 6-fructosyltransferase
3.1.1.3	Triacylglycerol lipase	Triacylglycerol acylhydrolase	Lipase, tributyrase, triglyceride lipase
3.1.1.11	Pectin esterase	Pectin pectylhydrolase	Pectin methylesterase, pectin (de)methoxylase
3.2.1.1	α-amylase	1,4-α-D-glucan glucanohydrolase	Glycogenase, dextrinogen amylase
3.2.1.2	ß-amylase	1,4-α-D-glucan maltohydrolase	Saccharogen amylase, glycogenase
3.2.1.3	Glucan 1,4-α-glucosidase	1,4-α-D-glucan glucohydrolase	Glucoamylase, amyloglucosidase, ɣ-amylase, α-glucosidase, exo-1,4-α-glucosidase
3.2.1.4	Cellulase	1,4-(1,3;1,4)-ß-D-glucan 4-glucanohydrolase	Endo-1,4-ß-glucanase
3.2.1.7	Inulinase	2,1-ß-D-fructan fructanohydrolase	Inulase
3.2.1.10	Oligo-1,6-glucosidase	Dextrin 6-α-D-glucanohydrolase	Limit dextrinase, isomaltase, sucrase-isomaltase
3.2.1.11	Dextranase	1,6-α-D-glucan 6-glucanohydrolase	
3.2.1.15	Polygalacturonase	Poly(1,4-α-D-galacturonide) glycanohydrolase	Pectin depolymerase, pectinase
3.2.1.23	ß-galactosidase	ß-D-galactoside galactohydrolase	Lactase
3.2.1.26	ß-fructofuranosidase	ß-D-fructofuranoside fructohydrolase	Invertase, saccharase
5.3.1.5	Xylose isomerase	D-xylose ketol-isomerase	Glucose isomerase
5.3.1.9	Glucose-6-phosphate isomerase	D-glucose-6-phosphate ketol-isomerase	Phosphohexose isomerase, glucoseisomerase

Classification and Numeration

Enzymes are divided into six classes according to the reaction catalyzed. The classes are further subdivided in the manner illustrated by a few examples in Table 5. The criteria upon which the subdivision depends necessarily differ from one class of enzymes to another. Oxidoreductases, for example, are subdivided according to the group on which the enzyme acts, whereas for transferases it is the group transferred that determines the subdivision to which the enzyme is allocated. The subdivisions are divided yet again into sub-subdivisions in which the enzymes are numbered consecutively.

Table 5. Classes and some subclasses of enzymes

Class, subclasses, specificity	Enzyme (example) with EC number
1. Oxidoreductases (redox reactions)	
1.1 Acting on =CH-OH	Glucose oxidase (EC 1.1.3.4)
1.2 Acting on =C=O	Formate dehydrogenase (EC 1.2.1.2)
1.3 Acting on =C=CH-	Fumarate reductase (EC 1.3.1.6)
1.4 Acting on =CH-NH$_2$	Glutamate dehydrogenase (EC 1.4.1.3)
2. Transferases (group transfer)	
2.1 C1-groups	Thiol methyltransferase (EC 2.1.1.9)
2.2 Aldehyde- or keto groups	Transaldolase (EC 2.2.1.2)
2.3 Acyl groups	Fatty-acid synthase (EC 2.3.1.85)
2.4 Glycosyl groups	Dextran sucrase (EC 2.4.1.5)
3. Hydrolases (hydrolytic reactions)	
3.1 Acting on ester bonds	Pectin esterase (EC 3.1.1.11)
3.2 Acting on glycosidic bonds	ß-Amylase (EC 3.2.1.2)
3.3 Acting on ether bonds	Ribosylhomocysteinase (EC 3.3.1.3)
3.4 Acting on peptide bonds	Papain (3.4.22.2)
4. Lyases (additions to double bonds)	
4.1 Acting on =C=C=	Pyruvate decarboxylase (EC 4.1.1.1)
4.2 Acting on =C=O	Pectate lyase (EC 4.2.2.2)
4.3 Acting on =C=N-	Argininosuccinate lyase (EC 4.3.2.1)
5. Isomerases (intramolecular changes)	
5.1 Racemases and epimerases	Glutamate racemase (EC 5.1.1.3)
5.2 Cis-trans-isomerases	Maleate isomerase (EC 5.2.1.1)
5.3 Intramolecular oxidoreductases	Xylose isomerase (EC 5.3.1.5)
5.4 Intramolecular transferases	Phosphoglycerate mutase (EC 5.4.2.1)
6. Ligases (bond-forming reactions)	
6.1 Forming C-O-bonds	Lysine-tRNA ligase (EC 6.1.1.6)
6.2 Forming C-S-bonds	Biotin-CoA ligase (EC 6.2.1.11)
6.3 Forming C-N-bonds	Glutathione synthase (EC 6.3.2.3)
6.4 Forming C-C-bonds	Pyruvate carboxylase (EC 6.4.1.1)

Every enzyme that is recognized as such by the Nomenclature Commission of the IUB receives a code number (EC number) consisting of four parts separated by dots. The code number for L-lactate dehydrogenase, for example, is 1.1.1.27. In this case the individual numbers have the following meaning:

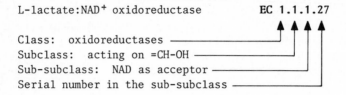

An enzyme code number is allocated only once, so that it provides an unambiguous and clear definition of the enzyme concerned. In publications, confusion can be avoided if the EC number is indicated in brackets after the first mention of the enzyme, e.g., ß-amylase (EC 3.2.1.2).

1.4 Definition and Classification of Immobilized Biocatalysts

A biocatalyst is termed "immobilized" if its mobility has been restricted by chemical or physical means. This (artificial) limitation of mobility (immobilization) may be achieved by widely differing methods, such as binding the biocatalysts to one another or to carrier substances by entrapping in the network of a polymer matrix or by membrane confinement. An essential criterion for defining a system as immobilized is that human interference has to be involved. In contrast naturally occurring enzymes, which may be bound to cellular or membrane structures, regardless of the fact that they are bound, are termed "native."

The definition of immobilization as the restriction of mobility by human action is not without its borderline problems, as is illustrated by the following example. Naturally occurring microorganisms such as some strains of brewer's yeast possess the ability to flocculate, i.e., due to certain properties of the cell wall the yeast cells clump together to produce floccules. There is no doubt that these are native cells, or native biocatalysts if the cells are considered as complex biocatalysts. This ability to flocculate can be "learnt" by nonflocculating yeast cells, or conferred by mutagenesis. The progeny of the mutated cells are able to flocculate and pass on this property to ensuing generations which can then be employed as complex biocatalysts in technical processes. In this book, such cases will be treated as immobilized systems.

A classification of immobilized enzymes was recommended by an international body for the first time in 1971 at the first Enzyme Engineering

Conference in Henniker (USA). The initial distinction was made between the two large groups of bound and encapsulated enzymes. The bound enzymes were then further subdivided into those bound by covalent linkages, and the encapsulated enzymes into matrix-trapped and microencapsulated. Although originally intended for immobilized enzymes, the classification can also be applied to more complex biocatalysts such as organelles or whole cells. Fig. 4 shows a considerably extended classification based on the initial suggestions.

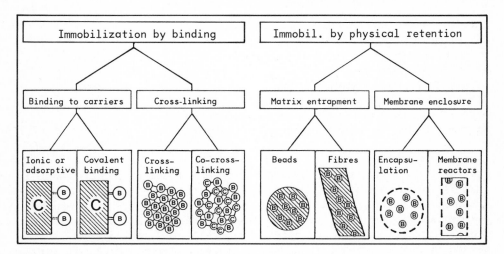

Fig. 4. Classification of immobilized biocatalysts;
  B biocatalyst unit; **C** carrier unit

A classification on the basis of the method of immobilization (cf. Fig. 4) covers the most important, although by no means all, known forms of immobilized biocatalysts, and other systems are in common use. In fact, the more recent developments in the field of immobilization involve combinations of types that can no longer be fitted into the simple scheme shown in Fig. 4. For this reason we often encounter other classifications, of which only two will be outlined here.

A distinction in common use is that between immobilized enzymes and immobilized whole cells, which can then, for example, be further subdivided into microorganisms, animal or plant cells, either living or dead. Borderline cases present difficulties whatever the method of classification. How, for example, are we to classify dead cells that contain only one desired enzyme (such as glucose isomerase)? Despite the fact that after immobilization the cell acts as an enzyme carrier, rendering enzyme isolation unnecessary, it is in fact a single enzyme that has assumed the biocatalytic function important for the process in question. Whether this is an immobilized cell or an immobilized enzyme is more or less a matter of personal point of view.

Another means of classification is presented by the different cofactor requirements of immobilized enzymes. In general, enzymes that require no cofactor are very simple to immobilize. The same is true of enzymes whose cofactor is tightly bound as a prosthetic group to the apoenzyme. Complications arise, however, if the coenzyme has to be regenerated by the action of a second enzyme. The situation becomes even more difficult in complex multienzyme systems requiring dissociating coenzymes. A classification on the basis of complexity is shown in Table 6.

Table 6. Classification according to complexity

Class	Description of the system
I	Enzyme without cofactor
II	Enzyme with prosthetic group
III	Enzyme with dissociating coenzyme
IV	Multienzyme system with dissociating coenzymes

The two first groups of systems shown in Table 6 are frequently known as immobilized enzymes of the first generation, those of groups III and IV as those of the second generation. The same classification can be employed for organelles and whole cells because, in each case, enzymes are the biocatalytically active (sub)units. Living cells or intact organelles with several active enzymes thus belong to group IV. A dead cell containing a single important enzyme, such as the example described earlier, belongs to group I, II or III, depending upon its cofactor requirements. Even a classification according to Table 6 is not feasible in every case, for example, as when it is difficult to judge whether a coenzyme is tightly bound to its apoenzyme or dissociates from it.

1.5 Reasons for Immobilization

Basically, immobilization of biocatalysts is undertaken either for the purpose of basic research or for use in technical processes of commercial interest. In the following brief outline of the reasons for immobilization and its chances of finding application, the emphasis is on the industrial aspects.

As already defined in Sect. 1.4, immobilization means that biocatalysts, while retaining their catalytic activity, are confined within a certain space or are bound to solid carriers or to one another. Instead of homogeneous catalysis, in which substrate and catalyst are present in a homogeneous solution, immobilization makes possible heterogeneous catalysis, which can be of very considerable advantage. With the technical methods in common use it is generally impossible to separate efficiently the dissolved or finely suspended native biocatalysts from the products of the bioconversion. This situation can be altered by the use of immobilized biocatalysts and, in addition, the process can be carried out continuously.

A vital criterion in industrial processes is a high yield per space and time, which can also be expressed as the volumetric productivity in kg product per cubic meter reactor volume and hour. Immobilization makes it possible to achieve and maintain a high biocatalytic activity in a small volume. This leads not only to a high volumetric productivity, but also to gentler conditions for substrates and products by reducing the time for which they are exposed to the reaction conditions.

Immobilized biocatalysts are always in competition with native biocatalysts. Which of the two forms is preferable can only be decided by careful consideration of the particular situation. In the case of very cheap, soluble, highly active enzymes, immobilization is often scarcely worthwhile. This is one of the reasons why immobilized hydrolases, for instance, have not so far come into common use in starch hydrolysis, one of the classical fields of enzyme application.

The nature of the substrate also plays a large role in the decision as to whether immobilized or native catalysts should be used. In leather processing or for tenderizing meat, for example, the use of immobilized enzymes is hardly feasible. Only if the substrate forms a clear solution and has a low molecular weight can immobilized enzymes be employed without the fear of problems arising. The reasons for this are that turbidity complicates the separation of immobilized biocatalysts from the substrate particles and that large substrate molecules, on account of their relative immobility, are more readily converted by native (mobile) than by immobilized enzymes.

For an immobilized catalyst, the chances of being chosen for a particular procedure depend, to a very large degree upon whether its properties are changed for the better as compared with those of its native form. At the same time attention has to be paid to the special requirements of the intended application. It is known that immobilization can cause a significant change in the stability, K_m value, pH and temperature relationships of biocatalysts. At present, however, the prediction of how and to what extent such changes occur is in a preliminary phase. Further research into the principles underlying the activation and stabilization mechanisms would also mean a great improvement in the prospects for the use of specially tailored immobilized biocatalysts in industry.

1.6 The History of Immobilization

Long before immobilization techniques as such were recognized, immobilized catalysts were already in use, even in large-scale processes (see Table 7). In about 1815 it was discovered by purely empirical means that vinegar could be efficiently produced by letting alcohol-containing solutions trickle over wood shavings. Of course it was not realized that this procedure caused the acetic acid bacteria to adhere to the shavings, which amounted to an effective immobilization. The development of a method from purely empirical beginnings is not at all unusual in biotechnology. In fact, many microbial procedures, such as the production of alcoholic beverages, sour bread and cheese, were practiced long before the microorganisms involved were known.

In Table 7 an attempt is made to trace the steps in the development of immobilization techniques, taking as a criterion the first large-scale use, rather than the laboratory development of the process. The biocatalysts allocated to step II are commonly designated as first generation, those listed under step III as second generation.

Table 7. Steps in the development of the immobilization technique

Step	Date	Description	Typical processes
I	1815	Empirical use without knowledge of details of immobilization	Trickling processes for acetic acid and waste water treatment
II	1969	Simple one-enzyme reactions without cofactor regeneration	Production of L-amino acids, Isomerization of glucose
III	1985	Two-enzyme reactions including cofactor regeneration	Production of L-amino acids in membrane reactors

As early as 1916 it was reported by Nelson and Griffin that active charcoal retained its ability to break down sucrose following contact with yeast invertase and subsequent washing. This is the first published report of enzyme immobilization. However, no further attention was paid to it from the point of view of immobilization and it was not until after World War II that further studies on the binding of enzymes appeared. In 1948 Sumner reported the immobilization of urease by treatment with alcohol and common salt. This was followed by a few publications in the 1950's during which time Grubhofer and Schleith, as well as Manecke and others, were able to demonstrate that specifically acting synthetic polymers could be used to bind physiologically active proteins.

From the 1960's onward the explosive increase in publications reflected the worldwide interest aroused by and still being taken in immobilization. An almost inevitable and logical consequence of the intensification of scientific interest in the field was the first industrial application in 1969. On the basis of the work done by I. Chibata, the Japanese company Tanabe Seiyaku introduced a method of producing L-amino acids from racemic mixtures using ionically-bound L-aminoacylase. Further large-scale processes employing simple enzyme reactions without cofactor regeneration followed in the early 1970's principally in the USA and Japan.

The first Enzyme Engineering Conference was held in the USA in 1971. It was predominantly concerned with immobilized enzymes, and similar meetings are still held every second year. Among other achievments, the term "immobilized enzymes" was recommended at the first meeting. Previous to this, a number of different names, such as "fixed," insolubilized," "matrix bound," etc. had been used instead of "immobilized." Another result of the same conference, i.e., the working out of a classification for immobilized enzymes, has already been mentioned in Sect. 1.4.

Whereas up to the 1970's only immobilized single enzymes were of interest for research and development, it was at about this time that more complex systems such as living cells began to receive attention; around the mid-1970's interest was further extended to include organelles. By about 1970 Mosbach and others had developed the first laboratory methods for binding and regenerating coenzymes. Then, at the end of the 1970's not only microorganisms but also cells from plant- and animal-tissue cultures were immobilized.

The latest milestone deserving mention is the introduction in the mid-1980's of the industrial production of L-amino acids from keto acids. This is a two-enzyme procedure involving coenzyme regeneration in membrane reactors (see Chap. 5) and was developed by the research teams of Wandrey and Kula in Germany. It is at present being employed on an industrial scale by the West German company Degussa. It thus seems that a breakthrough has been achieved in the technical application of immobilized biocatalysts of the second generation as well.

1.7 Economic Importance

It is not easy to obtain reliable marketing data on enzymes. Available publications on the subject are based on widely differing criteria for estimates and evaluations, with the result that the conclusions diverge widely from one another. Nevertheless, I have made an attempt to arrive at rough figures (shown in Table 8) for the worldwide total annual sales in US$ of some enzymes. The figures are based on published data, personal knowledge and information from colleagues. I have included only those enzymes that, with annual sales exceeding 0.5 million US$, can be said to be industrially important. The calculations are not based on the sales price of enzyme specialties, but on the much lower prices for bulk material.

Table 8. Estimations of the market volume of the most important enzymes

Enzyme group	Enzyme	Annual market volume in US$
Proteolytic enzymes	Bacterial protease	105 million
	Calf rennin	50 million
	Microbial rennin	18 million
	Papain	13 million
	Pancreas protease	7 million
	Fungal protease	6 million
	Pepsin	1 million
Enzymes degrading polysaccharides	Glucoamylase	55 million
	Bacterial α-amylase	25 million
	Pectinase	20 million
	Fungal α-amylase	5 million
	Cellulase and other ß-glucanases	2 million
Other enzymes	Glucose isomerase	23 million
	Invertase	3 million
	Glucose oxidase and catalase	2 million
	Lipase	1 million
	ß-Galactosidase	1 million
Total		US$ 340 million

Of the industrially important enzymes listed in Table 8, glucose isomerase is almost exclusively, lactase to a large extent, employed in the immobilized form. Other immobilized enzymes, such as penicillin acylase, L-amino acylase and others, scarcely appear on the international market. They are in many cases prepared "on the spot," and thanks to their stability the amounts required and the cost involved are very small in relation to the value of their end products. For biocatalysts of greater complexity than single enzymes there is at present no market worth mentioning.

Whereas the number of companies selling enzymes approaches a thousand, the number of enzyme producers is much smaller. Altogether in the USA and Western Europe there are only about 30 firms producing enzymes. Most enzyme producers are branches of the chemical-pharmaceutical industry, for whom the income from the enzymes plays a very minor role in their annual turnover. In Japan alone there are about 20 companies producing enzymes, some, however, producing solely for their own use. About 90 % of the world production of enzymes comes from the ten largest producers.

Of the branches of industry in which enzymes are employed, the washing powder and starch industries are by far the most important. Together they consume almost two-thirds of the world's total enzyme output. Figure 5 shows the distribution of enzymes on the international market; it reveals that the food industry, which includes the starch industry, accounts for the largest part. If it is considered that about 2500 enzymes are known at the present time, and at least part of them well characterized, the figures of 300 enzymes on the market and less than 20 in production on an industrial scale appear surprisingly low. Even the total market volume of approximately 350 million US$ annually is modest in comparison with the turnover of the large chemical and pharmaceutical concerns. It has to be remembered, however, that enzymes, in their role of specific catalysts make possible a large number of processes whose products have a market value many thousand times that of the enzymes themselves.

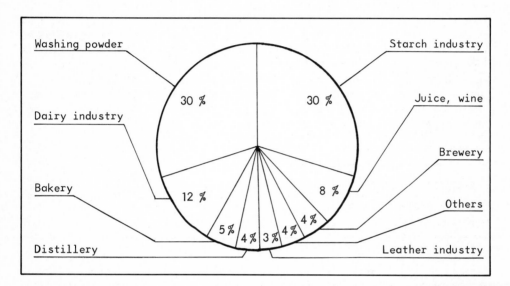

Fig. 5. The most important application fields of enzymes

In general, the role of enzymes, and particularly of immobilized enzymes in reactions that represent little hazard to either product or environment, seems to be rapidly gaining in importance in the food and chemical-pharmaceutical industries. The increasing interest need not necessarily be reflected in a noticeable increase in the sales of the enzymes themselves, but rather in that of the products. A major aim of intensified research and development in the fields of immobilization and stabilization is an ever-larger quantity of product from an ever-smaller amount of enzyme.

2 Methods of Immobilization

2.1 Adsorption

Adsorption is the simplest and the oldest method of immobilizing an enzyme onto a water-insoluble carrier. It has already been mentioned (p. 18) that as early as 1916 Nelson and Griffin observed that invertase adsorbed on active charcoal retained its sucrose-splitting activity. Since this discovery, the adsorption method has been applied for innumerable enzymes and whole cells. That whole cells can be made to adhere to suitable solid bodies has been known for even longer than the adsorption of single enzymes: in the nineteenth century (see Sect. 1.6) bacteria attached to wood shavings were already used in the production of vinegar.

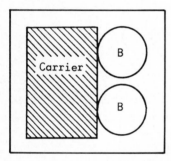

Fig. 6. Biocatalysts (**B**) bound to a carrier by adsorption

In adsorption, the biocatalysts are held to the surface of the carriers by physical forces (van der Waals forces). Frequently, however, additional forces are involved in the interaction between carrier and biocatalyst, principally hydrophobic interactions, hydrogen bridges and heteropolar (ionic) bonds, so that it is often impossible to arrive at a clear definition of the adhesion mechanisms of individual enzymes, and even more so of whole cells. In such cases we simply speak of "adsorption," at the same time fully realizing that considerably more than straightforward physical adsorption is involved.

Adsorbents can be selected from a wide variety of inorganic and organic substances, as well as synthetic polymers. A few examples of enzymes bound by adsorption and the adsorbents used are listed in Table 9.

The advantages of adsorption are that it is simple to carry out and has little influence on the conformation of the biocatalysts. In fact, to establish an adsorptive bond it suffices to bring a suitable adsorbent into contact with an aqueous solution or suspension of the biocatalyst for some time. Unphysiological coupling conditions or the use of chemicals harmful to enzymes or cells, such as are often employed in other methods of immobilization, are unnecessary in this procedure.

Table 9. Some biocatalysts immobilized by adsorption: Examples taken from the literature

Biocatalyst	Carrier	Reference
Alcohol dehydrogenase	Polyaminomethyl styrene	Schöpp and Grunow (1986)
Alkaline phosphatase	Aluminium	Koga et al. (1984)
Enniatine synthetase	Propylagarose	Madry et al. (1984)
Glucoamylase	Aluminium oxide	Krakowiak et al. (1984)
Glucoamylase	Titania-activated glass	Cabral et al. (1984)
Glucose oxidase	Activated charcoal	Miyawaki and Wingard (1983)
Glucanotransferase	Synthetic resin	Kato and Horikoshi (1983)
Phosphatase b	Butyl agarose	Jennissen (1986)
Acetobacter cells	Cellulose derivatives	Bar et al. (1986)
Clostridia	Cellulose, hemicellulose	Wiegel and Dykstra (1984)
Clostridia	Wood shavings	Förberg and Häggström (1984)
Pseudomonads	Activated charcoal	Ehrhardt and Rehm (1985)
Yeast cells	Charcoal pellets	Gianetto et al. (1986)
Yeast cells	Glass	Haecht et al. (1985)
Yeast cells	Stainless steel, polyester	Black et al. (1984)

A disadvantage, however, is the relative weakness of adsorptive binding forces. Adsorbed biocatalysts are easily desorbed by temperature fluctuations, and even more readily by changes in substrate and ionic concentrations. For this reason, particular attention must be paid to maintaining constant reaction conditions when employing biocatalysts immobilized by adsorption.

A marked improvement in yield of bound biocatalyst and in the durability of its binding can sometimes be attained by irradiating or by coating with transitional metals the surface intended for binding. Greater stability of binding has also been achieved by modifying the biocatalyst itself beforehand, e.g., by treating yeast cells with aluminum ions before binding them to glass. For the most part, in addition to physical adsorption, methods involving alteration of adsorbents and biocatalysts also involve electrostatic forces, as will be described in more detail in the following section.

2.2 Ionic Binding

Ionic or heteropolar binding, as shown schematically in Fig. 7, is based on the electrostatic attraction between oppositely charged groups of the carrier material and the biocatalysts. Suitable carriers are chiefly the more common commercially available ion exchangers based on polysaccharides or synthetic resins. The basic skeleton of an anion exchanger is positively charged, i.e., it bears anions (e.g., OH^--groups) which can be exchanged for other anions (e.g., negatively charged groups on enzymes). The situation is the reverse in cation exchangers (cf. Fig. 7); they can be used to bind the positively charged groups on biocatalysts.

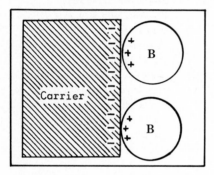

Fig. 7. Biocatalysts (**B**) ionically bound to a polyanionic carrier

Biocatalyst proteins are ampholytes, i.e., they have both acid and alkaline groups which, depending on the pH value of the surrounding medium, can be present uncharged or, by dissociation, in the negatively charged form or, by protonization, in the positively charged form. The carboxyl and amino groups are mainly responsible for the electric charge of proteins. At the isoelectric point the overall charge on a protein is to outward appearances neutral, i.e., in addition to the uncharged carboxyl and amino groups there are equal numbers of NH_3^+- and COO^--groups in the molecule. At pH values above the isoelectric point (IEP) the negatively charged carboxyl groups (COO^-) predominate because more NH_2 groups are present in the uncharged form. At lower pH values, below the IEP, the carboxyl groups are mainly present in the undissociated form (COOH) whereas the amino groups are protonized as NH_3^+. The overall charge on the protein is therefore positive. However, on account of the size of the enzyme protein it is not so much the charge on the entire molecule that is of interest, but rather the individual groups exposed on the surface of the molecule. This explains why, at one and the same pH value, some enzymes can be bound to both cation- and anion-exchangers.

The immobilization of biocatalysts by ionic binding to suitable carriers is as simple a process as physical adsorption, and as early as 1956 Mitz described a catalase bound to DEAE-cellulose. To obtain ionically bound preparations it is usually sufficient to stir the carrier particle for some time in a solution or suspension of the biocatalyst, or to let an aqueous solution of the enzyme flow over the carrier particles (e.g., in a column). Mainly anion exchangers, such as DEAE-cellulose and DEAE-Sephadex, are in common use as carrier materials (cf. Table 10). When, in the course of use, the carrier material becomes exhausted due to detachment and inactivation of the enzymes or cells, the ion exchanger can usually be regenerated. It is freed of the enzyme remnants and reloaded with active biocatalysts.

Table 10. Some ionically bound biocatalysts: Examples taken from the literature

Biocatalyst	Carrier	Reference
Aminoacylase	DEAE-cellulose	Tosa et al. (1967)
Aminoacylase	DEAE-Sephadex	Tosa et al. (1969)
Aldehyde oxidase	Octylamino-Sepharose 4B	Angelino et al. (1985)
Catalase	DEAE-cellulose	Mitz (1956)
Dextransucrase	DEAE-Sephadex	Ogino (1970)
Glucose oxidase	DEAE-Sephadex, DEAE-cellulose	Kühn et al. (1980)
Invertase	DEAE-cellulose	Suzuki et al. (1966)
Invertase	Amberlite IRA	Boudrant and Ceheftel (1975)
Invertase	Polyethylene vinylalcohol	Imai et al. (1986)
Lactate dehydrogenase	Octylamino-Sephadex	Hofstee (1973)
Nitrile hydratase	DEAE-cellulose	Fradet et al. (1985)
Azotobacter spec.	Cellex E (cellulose)	Diluccio and Kirwan (1984)
Mammalian cells	DEAE-Sephadex	Giard et al. (1979)

One of the main reasons why ionically bound enzymes were the first immobilized enzymes to be employed for large-scale purposes, is the extreme simplicity and mildness of the coupling procedure. Since the end of the 1960's they have been used in the production of pure L-amino acids from synthetically produced racemic mixtures of DL-amino acids (see sect. 5.2).

Although the heteropolar binding of biocatalysts to ion exchangers is tighter than purely physical adsorption, it is weaker than covalent binding and is susceptible to interference, especially from other ions. In using ionically bound biocatalysts special attention must be paid to maintaining the correct ionic strength and pH conditions in order to prevent their detachment.

2.3 Covalent Binding

In covalent (homopolar) binding the atoms are linked by means of shared electron pairs. This can be exploited to produce a tight association between one biocatalyst and another, or between a biocatalyst and a carrier. A single link, however, is insufficient for coupling larger biocatalyst units, e.g., organelles or whole cells: complex biocatalysts of this kind have to be bound at several sites. Covalent binding is usually employed for coupling enzymes but not whole cells.

A frequently encountered disadvantage of immobilization by covalent binding is that it places great stress on the enzymes. The necessary harshness of the immobilization procedure nearly always leads to considerable changes in conformation and a resultant loss of catalytic activity.

The α- and ϵ-amino groups, as well as carboxyl, sulfhydryl, hydroxyl, imidazol and phenolic groups of the biocatalyst protein can be used as functional groups for covalent binding. Some of these groups, e.g., SH and ϵ-amino groups, can react directly with appropriate groups of the carrier. Others, e.g., OH groups, have as a rule to be activated before they can react with a carrier group.

Activation of the group destined for binding is often performed on the carrier rather than on the enzyme protein, thus reducing the risk of diminishing the latter's catalytic activity. Some commonly employed methods for the activation of groups, which will be discussed in more detail further on, are cyanogen bromide (BrCN) activation of OH groups or chloride activation of COOH groups, which gives rise to reactive COCl groupings.

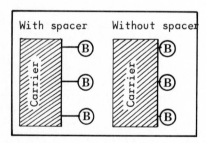

Fig. 8. Biocatalysts (B) covalently bound to a carrier with and without spacer

The connection between carrier and biocatalyst can be achieved either by direct linkage between the components or via an intercalated link of differing length, a so-called spacer. The spacer molecule gives a greater degree of mobility to the coupled biocatalyst so that its activity can, under certain circumstances, be higher than if it is bound directly to the carrier. Figure 8 shows schematically a biocatalyst joined directly and via a spacer molecule.

In the following, a few examples have been selected from the very large number of carrier materials and methods used for covalent binding.

Inorganic Carriers

Porous glass is the most commonly used inorganic carrier substance. It
contains a large number of OH groups on its surface. By treatment with
silanes, particularly aminoalkylethoxysilane or aminoalkylchlorosilane,
the surface of the glass can be activated, i.e., its ability to participate in further reactions is enhanced. Activation with aminoalkyl ethoxy-
silane proceeds according to the following equation:

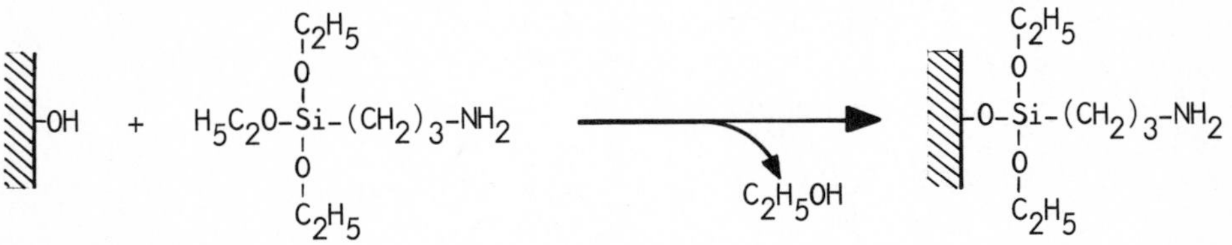

$$\text{glass}-OH \ + \ H_5C_2O-\underset{\underset{C_2H_5}{|}}{\overset{\overset{C_2H_5}{|}}{\overset{O}{\underset{O}{|}}}}Si-(CH_2)_3-NH_2 \ \longrightarrow \quad \xrightarrow{\quad C_2H_5OH} \quad \text{glass}-O-\underset{\underset{C_2H_5}{|}}{\overset{\overset{C_2H_5}{|}}{\overset{O}{\underset{O}{|}}}}Si-(CH_2)_3-NH_2$$

The glass is thus provided with alkyl amine groups and with the help of
bifunctional reagents such as dialdehydes or diisocyanates can be coupled
with enzymes. The example in Fig. 9 shows the coupling of an enzyme to
glass with the aid of glutaraldehyde, the most popular bifunctional
reagent for enzyme immobilization.

Another common procedure (Fig. 10, left) is the further treatment of
alkylamine glass with thiophosgene to give highly reactive isothiocyanate
groups (-N=C=S) to which the free amino acid groups of enzymes can be
bound. Figure 10 (right) illustrates how the p-aminobenzoyl derivative of
alkylamine glass can be produced by treatment with p-nitrobenzoylchloride.
Subsequent diazotization to the diazonium derivative gives a carrier that
readily couples with the amino groups of enzymes.

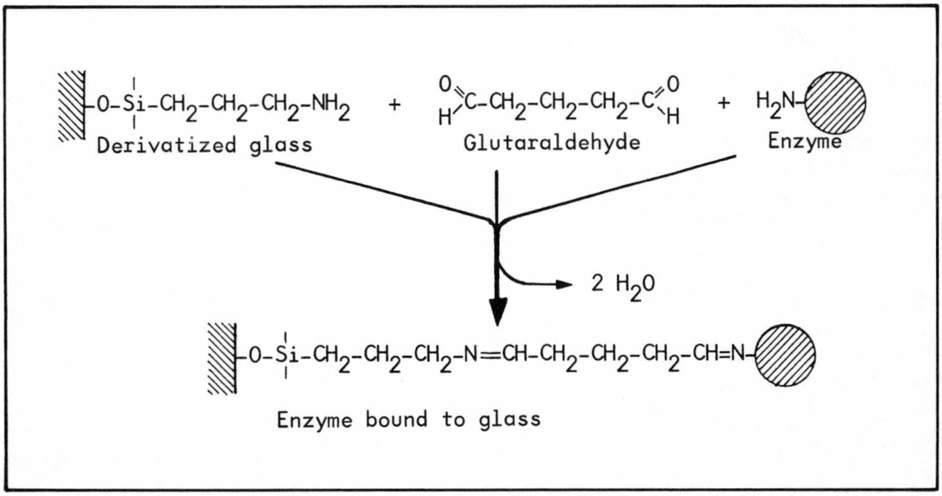

Fig. 9. Coupling of enzyme to glass by means of glutaraldehyde

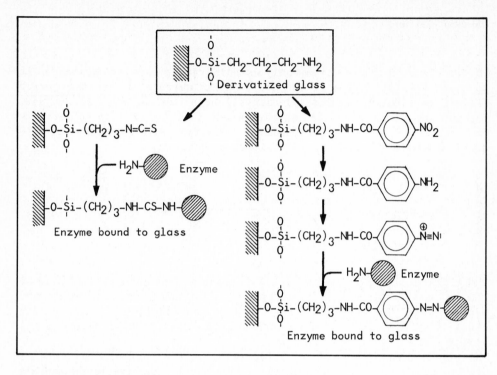

Fig. 10. Further derivatization of glass and covalent coupling of enzyme to the glass derivatives

In addition to glass, many other inorganic carrier substances can be used for the covalent coupling of single enzymes and whole cells. Some examples from the literature are listed in Table 11.

Table 11. Biocatalysts covalently bound to inorganic carriers

Biocatalyst	Carrier	Reference
Enzymes:		
Glucose oxidase	Porous glass	Hossain and Do (1985)
Papain	Porous glass	Weetall and Mason (1973)
Invertase	Porous kieselguhr	Monsan et al. (1984)
Whole cells:		
Enterobacteria	Borosilicate glass	Messing and Oppermann (1979)
Methanobacteria	Silochrome	Romanovskaya et al. (1981)
Yeast cells	Porous kieselguhr	Navarro and Durand (1977)
Yeast cells	Zirconium earth	Messing and Oppermann (1979)

Another method for the covalent binding of biocatalysts to inorganic carriers requires as a preliminary step the binding of natural or synthetic polymers to the carrier. In this way the inorganic carrier is surrounded by a polymer layer onto which biocatalysts can be coupled in a subsequent reaction. Further information about the covalent binding of biocatalysts to organic polymers is given in the following sections.

Natural Polymers as Carriers

Carriers based on natural polymers, such as cellulose, dextran, starch or agarose, all of them polysaccharides or polyuronides, enjoy great popularity in the field of biocatalyst immobilization. Agarose and dextran are usually subjected to additional cross-linking with epichlorhydrine to increase their stability. The cross-linked products, as Sephadex or Sepharose, are widely used in gel chromatography on account of their well-defined pore diameter. Natural polymers have in common that they bear large numbers of OH groups, although the reactivity of these groups is insufficient for direct coupling with enzyme proteins.

    Activation of the OH groups can be effected with cyanogen bromide (CNBr). In addition to other reaction products, this method gives rise to extremely reactive imido carbonates according to the following equation:

In a subsequent coupling reaction the activated carrier can react with the free amino groups of enzymes or other protein as shown below:

Carrier (act.) Enzyme   Bound enzyme   Ammonia

The imido carbonates are sometimes further converted into iso-ureates or carbamates and if necessary provided with other reactive groups before coupling with enzymes.

    CNBr-activated derivatives of natural polymers are among the most commonly used carriers in enzyme immobilization and are available commercially. In addition to the CNBr method, many other means of activation are known. Figure 11 shows as an example the starch derivative, dialdehyde starch, whose aldehyde groups can readily be used for further reactions and thus for enzyme coupling. In the case of cellulose a popular modification consists in allowing the substance to react with chloracetic acid to produce carboxymethyl cellulose. This can be converted to the corresponding azide, in which form it readily reacts with enzyme proteins.

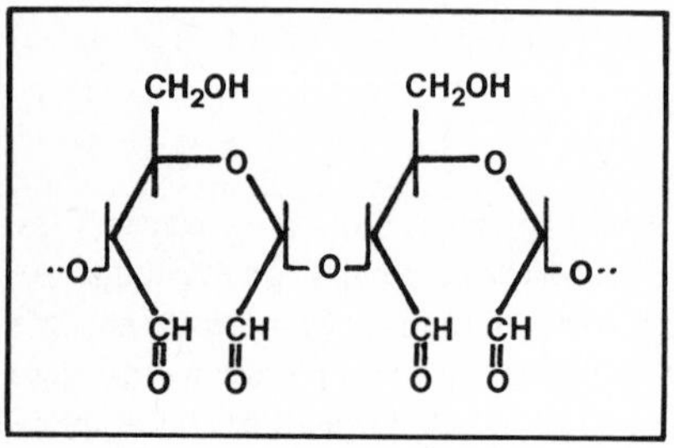

Fig. 11. Part of a dialdehyde starch chain

Covalent coupling to natural and other polymers is rarely used for the immobilizaton of whole cells because the reaction conditions are usually such that the cells can seldom be kept alive. Even in the example quoted in Table 12, where cells of *Bacillus subtilis* were bound to agarose, the viability of the cells could not be preserved so that this is not, strictly speaking, a case of bound whole cells, but rather of intracellular enzymes retained within the bound cell structure.

Table 12. Biocatalysts covalently bound to natural polymers

Biocatalyst	Carrier	Reference
Enzymes:		
Chymotrypsin	Sepharose	Clark and Bailey (1984)
Dihydrofolate reductase	Sepharose	Ahmed and Dunlap (1984)
Epoxide hydrolase	Dextran	Ibrahim et al. (1985)
Invertase	Corn stover	Monsan and Combes (1984)
ß–Lactamase	Sepharose	Pastorino et al. (1986)
Lactate oxidase	Cellulose	Cannon et al. (1984)
Whole cells:		
Azotobacter species	Cellulose	Gainer et al. (1980)
Bacillus subtilis	Agarose	Chipley (1974)
Micrococcus luteus	CM–cellulose	Jack and Zajic (1977)
Saccharomyces cerevisiae	Cellulose	Okita et al. (1985)

Synthetic Polymers as Carriers

Synthetic polymers were already being employed for coupling biologically active materials in the 1950's, especially by the research teams headed by Isliker, Grubhofer and Manecke. Synthetic procedures, initially used for binding antigens, were soon also being used for enzyme immobilization. Over the course of the next decades great numbers of fully synthetic carriers were developed, more or less explicitly for the purpose of immobilizing biocatalysts.

The development of synthetic carriers has proceeded along two different lines. The first of these is based on the use of already available polymers that were initially developed for purposes other than enzyme immobilization. Just as with natural polymers (see pp. 29f), these substances are then provided with reactive groups for enzyme coupling. The second line of approach is to produce copolymers specifically for immobilization purposes, starting from the appropriate monomers and comonomers.

Many reactive copolymers have been produced, chiefly on the basis of acrylic and methacrylic acid derivatives. In some cases the polymers bear reactive fluorodinitrophenyl or isothiocyanate groups. Acrylate or methacrylate carriers with reactive oxiran or acid anhydride groups are also widely used and are very well suited for direct coupling with enzymes. Figure 12 shows the coupling of an enzyme to a carrier bearing oxiran groups. Carrier substances of this type are produced by Röhm GmbH (Darmstadt, W. Germany) on the basis of polyacrylamide and sold under the trade name Eupergit[R].

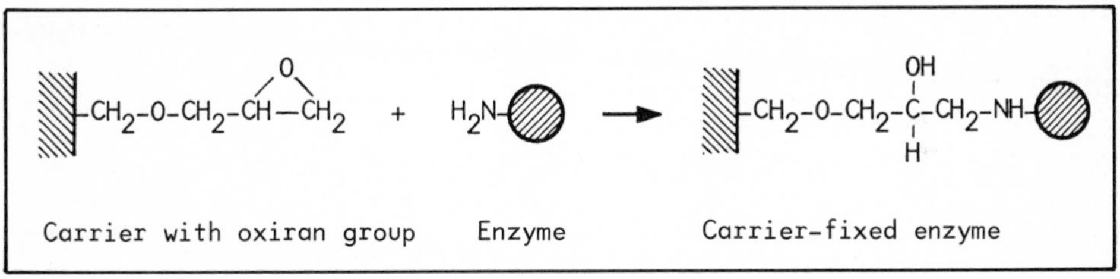

Fig. 12. Coupling of an enzyme to a carrier bearing oxiran groups

Another method of activation, chosen from the great variety of possibilities, is chloride activation, as depicted in Fig. 13. Resins bearing carboxyl groups, like the commercially available Amberlite IRC-50, can be activated in this way with thionylchloride. The carboxy resins thus produced have reactive acid chloride groups which then react with the free amino groups of enzymes to give peptide bonds (-CO-NH-) as shown in Fig. 13.

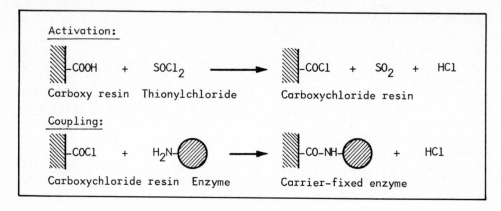

Fig. 13. Chloride activation of a carrier and coupling of enzyme thereto

Like all methods involving covalent bonds, those in which synthetic polymers are used as carriers have been mainly practiced with single enzymes and only rarely using whole cells as biocatalysts. The enormous variety of enzymes on the one hand and of polymer substances on the other is barely indicated by the few examples given in Table 13. Further examples and more details on the subject of carrier substances and methods for covalent coupling can be found in the literature listed at the end of this book.

Table 13. Biocatalysts covalently coupled to synthetic polymers

Biocatalyst	Carrier	Reference
Enzymes:		
Carboxypeptidase	Styrene-maleic anhydride copolymer	Kubo et al. (1986)
ß-Galactosidase	Nylon-acrylate copolymer	Beddows et al. (1981)
Invertase	Glycidylmethacrylate	Marek et al. (1984)
Trypsin	Polyacrylate-polyethylene copolymer	Beddows et al. (1982)
Tyrosinase	Polyacrylamide	Vilanova et al. (1984)
Urease	Methacrylate-acrylate copolymer	Raghunath et al. (1984)
Urease	Nylon, modified	Miyama et al. (1984)
Urease	Polymethylglutamate membrane	Dua et al. (1985)
Whole cells:		
Bacterial cells	Ethylene-maleic anhydride copolymer	Shimizu et al. (1975)
Yeast cells	Hydroxyalkylmethacrylate	Jirku et al. (1980)

2.4 Cross-linking

In the process of cross-linking, the individual biocatalytic units (en-zymes, organelles, whole cells) are joined to one another with the help of bi- or multi-functional reagents. In this way, as shown schematically in Fig. 14, very high-molecular, typically insoluble aggregates are formed.

Also depicted in Fig. 14 is the so-called co-cross-linking, in which, besides the catalytically active components (**B**), inactive molecules (**C**) are incorporated in the high-polymer network in order to improve the mechanical and enzymatic properties of the immobilized preparation.

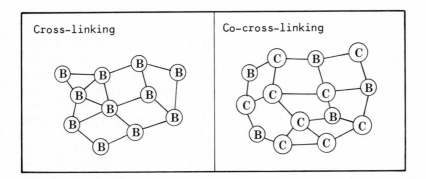

Fig. 14. Biocatalysts (**B**) immobilized by means of cross-linking and co-cross-linking

Cross-linking is a relatively simple process. One disadvantage of the particles produced in this way, however, is that they are usually gela-tinous and not particularly firm. This means that cross-linked biocata-lysts are less useful for packed beds (cf. Sect. 4.3). Another disadvan-tage is that many of the active biocatalyst units are bound within the particles formed by cross-linkage. Especially when high-molecular sub-strates are employed, and if the substrate concentration is low, access of the substrate to the innermost catalytic sites is limited by the unfa-vorable conditions for diffusion.

Since cross-linking and co-cross-linking usually involve bonds of the covalent kind (see Sect. 2.3), biocatalysts immobilized in this way fre-quently undergo changes in conformation with a resultant loss of activity.

By far the most commonly employed bifunctional reagent for cross-linking is glutardialdehyde, often simply called glutaraldehyde. The reac-tive aldehyde groups at the two ends of the glutaraldehyde react with free amino groups (ϵ-amino groups, N-terminal amino groups) of enzymes or of whole cells or parts of cells. As Fig. 15 shows, this leads to substances of the type of Schiff's bases.

CHO   NH$_2$ —◯   CH=N—◯
|
CH$_2$ CH$_2$
|
+   CH$_2$ +   ↘   CH$_2$
| 2 H$_2$O |
CH$_2$ CH$_2$
|
◯—NH$_2$ CHO   ◯—N=CH

Enzyme   + Glutaraldehyde + Enzyme ⟶ Linked enzymes

Fig. 15. Cross-linking of enzymes with glutaraldehyde

Biocatalysts cross-linked with glutaraldehyde have gained industrial importance in the isomerization of glucose (cf. Sect. 5.3). Some of the
immobilized preparations used in these large-scale processes are produced
simply by glutaraldehyde treatment of bacterial cell masses that have
formed fine particles. Also used for industrial purposes are glucose
isomerase preparations produced by co-cross-linking with gelatine: here,
too, glutaraldehyde is the linking agent.

In addition to glutaraldehyde, diisocyanates such as hexamethylene
diisocyanate and toluene diisocyanate are often used as linking agents. As
Fig. 16 shows, they form peptide bonds (-CO-NH-) with the biocatalyst
protein. Diisothiocyanate, which bears two -N=C=S groups, reacts in a
similar manner to the diisocyanate and binds to the free amino groups of
an enzyme by means of -NH-CS-NH-groups (cf. Fig. 10).

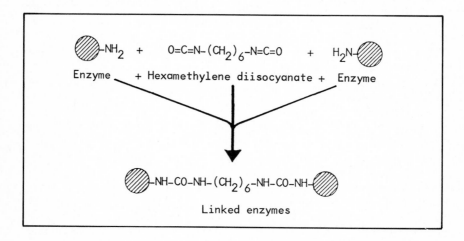

Fig. 16. Cross-linking of enzymes with hexamethylene diisocyanate

Cross-linking, as shown above, mostly involves the amino groups of the biocatalyst to be immobilized. Nevertheless, in occasional cases the sulfhydryl groups of cysteine, phenolic OH groups of tyrosine, or the imidazol group of histidine can also be used for binding. For example bisdiazobenzidine, as a bifunctional linking agent, reacts not only with the free amino groups, but also with the phenolic OH groups and imidazol groups of the enzyme protein.

Linking reagents with more than two functional groups are seldom used. With increasing molecular size and increasing number of reactive groups the reagent can gradually be regarded as a carrier with functional groups. A high-molecular dialdehyde starch (see Fig. 11), for example, is normally regarded as a carrier with reactive aldehyde groups. On the other hand, short-chain (low-molecular) fragments of the same dialdehyde starch can also be regarded as polyfunctional reagents.

Even without the participation of bi- or polyfunctional reagents, enzymes can be cross-linked through covalent disulfide bridges or peptide bonds. In addition, the weaker hydrogen bridges, hydrophobic interactions or van der Waals forces can be used in the direct binding of enzymes or whole cells to one another. Biocatalytically active aggregates (e.g., flocculating cells) arising in this way can be justifiably termed cross-linked biocatalysts.

A few examples of enzymes immobilized by cross-linking have been chosen from the very large number of such preparations and are presented in Table 14, together with the relevant literature.

Table 14. Examples of cross-linked biocatalysts

Biocatalyst	Cross-linking method	Reference
Bacterial cells	Co-cross-linking with egg albumin by glutaraldehyde	De Rosa et al. (1981)
Chymotrypsin	Formation of peptide bonds between enzyme molecules	Talsky and Gianit-sopoulos (1984)
ß-galactosidase	Co-cross-linking with egg albumin by glutaraldehyde	Kaul et al. (1984)
Glucose isomerase	Co-cross-linking with gelatine by glutaraldehyde	Bachmann et al. (1981)
Inulinase	Cross-linking of the enzyme-containing cells by glutaraldehyde	Workman and Day (1983)
Invertase	Cross-linking by adipinic acid dihydrazide	Barbaric et al. (1984)
Pepsin	Cross-linking of the enzyme by glutaraldehyde	Khan and Siddiqi (1984)
Trypsin	Co-cross-linking with human serum albumin by glutaraldehyde	Cohenford et al. (1986)

2.5 Matrix Entrapment

By matrix entrapment we mean that the enzymes or complex biocatalysts are embedded in natural or synthetic polymers, mostly of a gel-like structure. In order for the entrapped enzyme to fulfill its catalytic function it is essential that the substrate(s) and product(s) of the reaction are able to traverse the matrix. At the same time, the pores of the matrix should not be so large that the biocatalyst itself can escape.

The external form of the matrix-entrapped biocatalyst can be varied within wide limits to meet the specific requirements of the intended application. Thus, spherical, cylindrical, fiber- or sheet-like forms are possible. Figure 17 shows the most common, spherical, form as well as the fiber form.

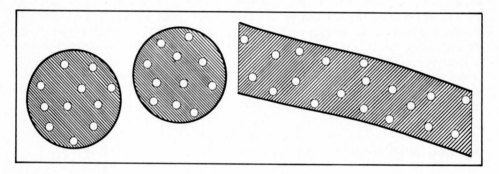

Fig. 17. Matrix-entrapped biocatalysts in spherical and fiber form

The chemical nature of the matrix materials, as well as the principles involved in their gelation can differ very widely. The five most important hardening principles are shown in Table 15: for each method of hardening two examples of matrix materials are given.

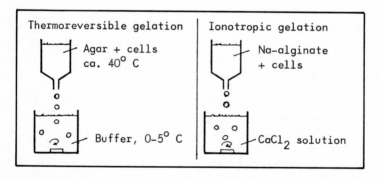

Fig. 18. Procedures for entrapping whole cells

Matrix entrapment is extremely popular for the immobilization of whole cells and active cell components (organelles). Particularly suitable matrix materials are the natural polymers alginate, carrageenan, pectin, agar and gelatine since they are nontoxic and the methods used for their gelation are very mild. These substances are not suitable for the entrapment of enzymes because, without further linking procedures, their mesh diameter is too large to retain single enzyme molecules.

The commonly employed method for entrapping whole cells in polymer spheres, as shown in Fig. 18, begins with the mixing of the gelatable protein with the cells destined for entrapment. In the cases of gelatine and agar as polymer, the substance is first dissolved in water at about 95° C, cooled to 40° C, then mixed with the cells and the mixture finally allowed to drop into ice-cold buffer solution for gelation. If alginate, pectin or carrageenan are chosen as matrix substances the water-soluble form (e.g., sodium alginate) is dissolved in water at room temperature and then mixed with the cells. Ionotropic gel formation ensues as soon as the mixture of alginate and cells is dropped or sprayed into a $CaCl_2$ solution. In the case of k-carrageenan, KCl is used instead of $CaCl_2$ for the ionotropic gelation of the polymer.

Table 15. Methods for entrapping biocatalysts

Hardening principle	Examples of matrix materials	Main applications
Thermoreversible gelation	Agar, gelatine	Living cells and cell organelles
Ionotropic gelation	Alginate, carrageenan	Living cells and cell organelles
Solvent precipitation	Cellulose acetate, polystyrene	Enzymes and dead cells
Polycondensation	Epoxy resins, polyurethane	Enzymes and dead cells
Polymerization	Polyacrylamide, -methacrylamide	Enzymes and dead cells

The natural polymers that are used for entrapment of biocatalysts usually lead to relatively soft products. Subsequent hardening procedures, such as treatment with glutaraldehyde, hexamethylenediamine or others, are of very limited applicability due to their detrimental effects on the cells. Simple drying (e.g., of the Ca-alginate spheres) using moderate temperatures under normal pressure or under vacuum is not harmful to the cells. The beads shrink in the drying process and even when rehydrated they do not regain their original wet volume. They remain harder than untreated but otherwise identical spheres.

A further disadvantage of the immobilized preparations based on ionotropic gel formation is their instability in the presence of ions with charges opposite to those required for the formation of the gel. For example, in the presence of excess sodium ions the "hardening" Ca ions in the alginate are displaced due to competition between Na and Ca ions for the alginate anion. Phosphate ions can also be responsible for a gradual disintegration of a calcium alginate sphere. Phosphate (or citrate), often used as buffer substances, also react with calcium, which is thus withdrawn from the alginate.

Polymers synthetically produced by polycondensation or polymerization are not particularly suitable for matrix entrapment of living organisms since the monomers employed are usually highly toxic and kill the cells. However, synthetic materials are frequently used for entrapping enzymes; polyacrylamide, in particular, is a very popular material for this purpose. The network of the polyacrylamide can be made dense enough to retain the enzyme molecules. Figure 19 shows an enzyme entrapped in polyacrylamide produced from the monomers acrylamide and bis-(N,N)-methylenebisacrylamide (BIS).

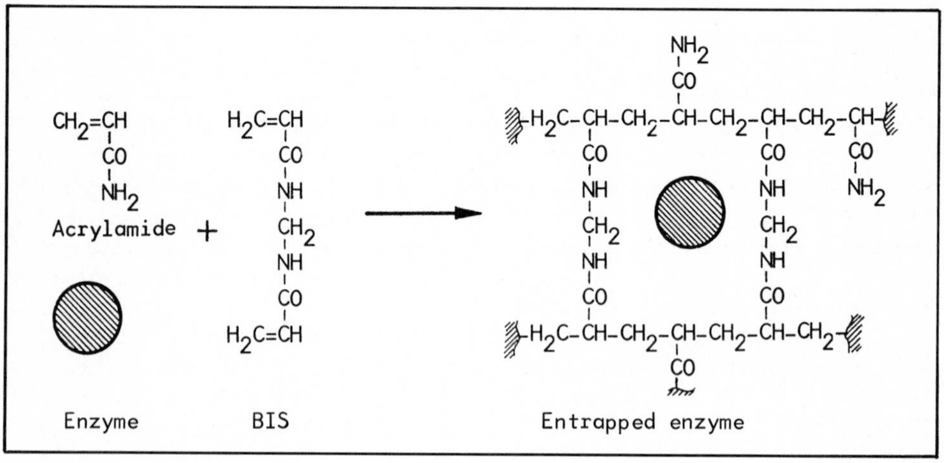

Fig. 19. Enzyme entrapment in polyacrylamide

Enzymes spun into fibers have even attained a certain degree of commercial importance (see Sect. 5.7). Figure 20 shows the principles involved in spinning enzymes into fibrous proteins. An aqueous solution of the enzyme is finely emulsified in a solution of the polymer in an organic solvent not miscible with water (e.g., cellulose triacetate in methylene chloride). The emulsion is then forced under pressure through a nozzle (internal diameter usually about 100 μm) into a polymer-precipitating solvent (e.g., toluene). The rapidly hardening fibers are drawn out of the nozzle and wound onto a spool. The final step is a thorough washing of the fibers with buffer before use.

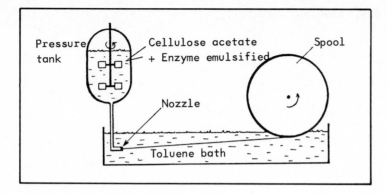

Fig. 20. Entrapment of enzymes by spinning with cellulose acetate

A very small selection from the vast field of matrix-entrapped bio-catalysts is given in Table 16, which is representative only inasmuch as almost exclusively whole cells and very rarely individual enzymes are immobilized by matrix entrapment.

Table 16. Examples of matrix-entrapped biocatalysts

Biocatalyst	Matrix material	Reference
Aminoacylase	Polyacrylamide	Mori et al. (1972)
Bromoperoxidase	Photo cross-linked resins	Itoh et al. (1987)
Bacterial cells	Photo cross-linked resins	Mazumder et al. (1985)
Bacterial cells	Polyacrylamide hydrazide	Bettmann and Rehm (1984)
Bacterial cells	Carrageenan	Umemura et al. (1984)
Bacterial protoplasts	Agar-acetylcellulose	Karube et al. (1985)
Cyanobacteria	Cross-linked albumin	Papageorgiou and Lagoyanni (1986)
Cyanobacteria	Polyvinyl, polyurethane foam	Brouers and Hall (1986)
Methanogenic bacteria	Agar	Dwyer et al. (1986)
Myxobacteria	Carrageenan	Younes et al.(1987)
Mold mycelium	Polyacrylamide	Bihari et al. (1984)
Mold mycelium	Carrageenan	Deo and Gaucher (1984)
Mold mycelium	Alginate	Eikmeier and Rehm (1984)
Nitrobacteria	Alginate	Tsai et al. (1986)
Plant cells	Alginate, Agarose	Vogel and Brodelius (1984)
Plant cells	Agar, Alginate, Carrageenan	Nakajima et al. (1985)
Yeast cells	Agarose	Shankar et al. (1985)
Yeast cells	Hydroxyethyl methacrylate	Cantarella et al. (1984)
Yeast cells	Alginate	Qureshi and Tamhane (1985)

2.6 Membrane Confinement

According to the definition in Sect. 1.4 (see p. 14), biocatalysts that are not bound to carriers but are in some way spatially restricted also come under the heading of immobilized catalysts.

Membranes can be used to delimit the space in which a biocatalyst exerts its activity. Although enzymes are in aqueous solution and cells in aqueous suspension, the membrane sets a limit to the space in which the reactions can proceed. In this respect the method imitates the situation met with in nature, since living cells also to a large extent confine their enzymes by means of an enveloping membrane, thus providing the conditions necessary for repeated or continuous operation of the enzyme pool.

Membrane confinement can be achieved by one of three main methods: by microencapsulation, by the liposome technique, and by using the biocatalysts in membrane reactors. All three methods result in the retention of the biocatalyst within a defined space by a semipermeable membrane which can be crossed by the substrate(s) and product(s) but is impermeable to the biocatalyst(s).

Microencapsulation

The method of microencapsulation has not yet been used for whole cells and only relatively seldom for enzymes. As shown in Fig. 21 the enzymes in aqueous solution are surrounded by a polymer membrane that permits free passage of product and substrate.

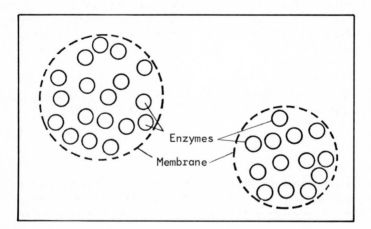

Fig. 21. Microencapsulated enzymes

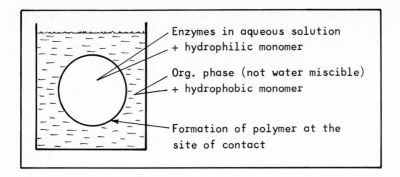

Fig. 22. Enzyme-encapsulation by boundary layer polymerization

Microencapsulation can, as illustrated in Fig. 22, be carried out as so-called boundary layer polymerization. In this method the enzymes and a hydrophilic monomer are emulsified to the desired droplet size (usually between 1 and 100 μm) in an organic solvent (e.g., cyclohexane or chloroform) that is not miscible with water. The size of the droplets can be influenced by the intensity of emulsification and by the addition of surface reactants. A hydrophobic monomer, which only dissolves in the solvent phase, is now added. The hydrophilic and hydrophobic monomers react at their site of contact - i.e., the boundary between aqueous and organic phase - to form a polymer. As a final step the capsules must be repeatedly washed to remove all traces of monomer.

In addition to di- and polyamines, glycols and polyphenols are widely used as hydrophilic monomers in boundary-layer polymerization. The hydrophobic monomers employed are bi- or polybasic acid chlorides, bishalogen formiate, and di- and polyisocyanates. The polymers forming at the phase boundary are polyamides, polyurethanes, polyesters or polyureas, depending on the monomers used.

To produce nylon (polyamide) microcapsules hexamethylene diamine is used as the hydrophilic monomer and sebacoyl chloride (sebacic acid dichloride) as the hydrophobic monomer. As shown below, the amino groups of the diamine react with the acid chloride group of the sebacoyl chloride and HCl is liberated:

$$H_2N-(CH_2)_6-NH_2 \ + \ ClOC-(CH_2)_8-COCl \ \longrightarrow \ H_2N-(CH_2)_6-NH-CO-(CH_2)_8-COCl$$
$$HCl$$

Since hexamethylene diamine and sebacoyl chloride have two reactive amino and two reactive acid chloride groups respectively, they join up to form long fibrous molecules which, when sufficiently densely packed in crisscrossed layers, constitute the microcapsule.

A disadvantage connected with boundary layer polymerization is that the enzyme and the aqueously dissolved monomers come into contact with one another, which often leads to partial inactivation of the catalyst. Enzyme damage from monomers is avoided in the so-called liquid drying method by using ready-made polymers which can be dissolved in an organic solvent immiscible with water. As indicated in Fig. 23, the enzyme in aqueous solution is then emulsified in this organic phase. To this first emulsion is added a larger amount of a so-called protective colloid (e.g., gelatine or albumin solution). Under suitable conditions of stirring and mixing, a second emulsion consisting of three phases is formed: inside is the enzyme in aqueous solution, surrounded by the solvent phase with the dissolved polymer, and on the outside the protective colloid in its aqueous solution. The organic solvent, which must have a lower boiling point than water, is evaporated under vacuum, in the course of which the polymer rehardens and now surrounds the enzyme solution as a microcapsule. The polymers used in the liquid drying method must be soluble in the organic solvents that are immiscible with water. Examples of such polymers are ethyl cellulose and polystyrol. Cyclohexane and chloroform are suitable solvents, both of them having a boiling point lower than that of water.

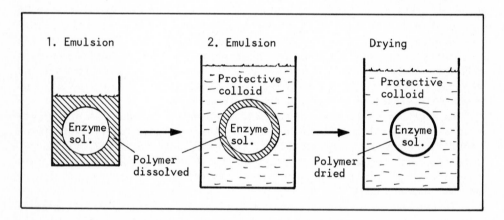

Fig. 23. Microencapsulation by the liquid drying method

Apart from the advantage that the use of enzyme-damaging monomers is avoided by liquid drying, the method also has a few disadvantages. Firstly, only relatively large microcapsules (roughly 20 μm in diameter) can be produced and, secondly, during the second emulsification the organic phase cannot always be made to surround the enzyme droplets in the desired manner shown in Fig. 23.

Another method for microencapsulation, the coacervation method or phase separation method, like the liquid drying method, also starts with the emulsification of an aqueous solution of the enzyme in an organic solution of the polymer. This is followed by the addition of a second organic solvent - also immiscible with water but miscible with the first organic solvent - which precipitates the polymer. A phase separation, also called coacervation, takes place within the organic phase. Under favorable

experimental conditions the polymer-containing phase collects in sufficient density around the enzyme droplets for the polymer coating to
fulfill the requirements of a microcapsule.

Liposome Technique

In contrast to the firm microcapsules produced in the methods described
above, the liposome technique is used to form soft, deformable and almost
liquid lipid membranes, comparable with those of living cells. They consist, typically, as illustrated in Fig. 24, of a lipid double layer en-
closing the enzymes or other substances in aqueous solution.

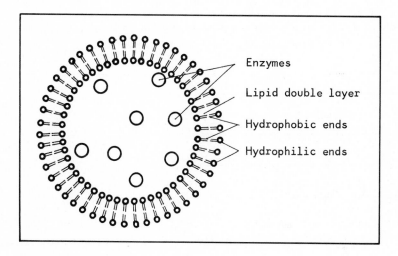

Fig. 24. Scheme of an enzyme-containing liposome

Liposomes can be formed, for example, by sonification (ultrasonic treatment) of phospholipids or other suitable lipids in aqueous solution. The
lipids used are amphilic, i.e., their chains have a hydrophilic and a
hydrophobic end. Such lipids tend to arrange themselves with their hydrophobic tails end to end, forming a lipid double layer with a hydrophobic
interior. The hydrophilic heads face outward (cf. Fig. 24). The double
layers have the tendency to merge with one another at their edges to form
closed vesicles which are called liposomes.
　　Since detergents are amphilic substances, they interfere with the
formation of lipid membranes so that their controlled addition can be used
to regulate the size of the liposome. Also based on this effect, another
method often used for making liposomes is to disperse suitable phospholipids (e.g., egg lecithin) in aqueous solution in the presence of a
detergent, which must then be removed by dialysis after the liposomes have
formed. It is also common practice to produce smaller liposomes from

larger ones by sonification. Liposomes of only a few nm diameter can be obtained in this way.

On account of the sensitivity of the double membrane structure, enzymes encapsulated in liposomes are obviously unsuited for harsh industrial conditions. However, they could be useful tools in medical therapy and in modelling natural systems for basic research.

Enzymes in Membrane Reactors

Yet another variation of membrane confinement is to let the biocatalysts perform their work in membrane reactors. They are retained in the reactor in hollow-fiber membranes or in sheet-like ultrafilter membranes and are in this way continuously available over longer periods of time. As the example in Fig. 25, in this case a hollow-fiber membrane, shows the reaction products can pass through the membrane and thus be removed continuously while the biocatalysts are held back by the membrane.

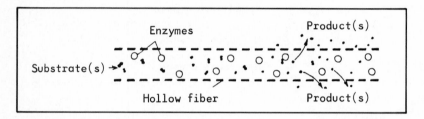

Fig. 25. Enzyme application in a hollow-fiber reactor

The two great advantages of this method are that commercially available and relatively cheap ultrafilter devices can be used, and that the biocatalysts are not exposed to inactivating steps. More detailed information is given in Sect. 4.4.

2.7 Combined Methods

In addition to the basic methods of immobilization described in the preceding Sects. 2.1 to 2.6, the scientific - and patent - literature contain
numerous combined methods and modifications, of which only a few are
considered here.

Some of the advantages and disadvantages of these methods have already
been mentioned in reviewing the individual methods of immobilization. The
chief goal in combining methods is to avoid every disadvantage of the
individual method but to exploit every advantage. Almost always, the aim
is to produce immobilized biocatalysts with the greatest possible specific
activity and stability.

Combination of Adsorption and Cross-linking

A simple combined method involving adsorption and covalent binding is
shown in Fig. 26. The enzymes are first adsorbed onto a carrier (e.g.,
silica gel or polyamide). In a second step the adsorbed enzyme molecules
are linked to one another by a bifunctional reagent (glutaraldehyde).

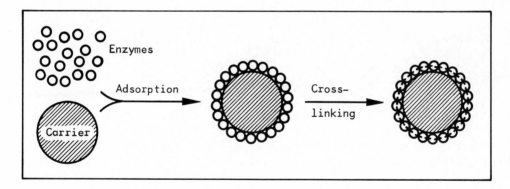

Fig. 26. Adsorption and subsequent cross-linking of enzymes according to
Hanes and Walsh (U.S. Pat. 3.796.634, 12.03.1974)

The resulting product, shown in Fig. 26, has the advantage that the enzyme
molecules are more stably bound than if they are only adsorbed. Reduced
diffusion of the substrate to the enzymes, such as encountered in normally
cross-linked enzymes, are seldom experienced with this method. Nevertheless, the disadvantage remains that only a relatively small amount of
enzyme - and consequently little activity - is bound per carrier mass,
since only the surface of the spherical carrier is available for binding.

Combined Methods Using Porous Carriers

Porous carrier materials are often used for the purpose of immobilization. Their large specific surface area (area per weight, or area per volume) as compared to nonporous carriers permits a greater loading with biocatalysts, which results in a much higher specific activity (activity per weight or activity per volume). As compared with crosslinked biocatalysts, given an adequate pore diameter of the carrier, access to the biocatalysts bound inside the carrier is easier.

Coupling of biocatalysts to porous carriers can be achieved by one of the basic methods described in Sects. 2.1 to 2.4. The covalent binding of enzymes to porous glass is described on p. 27. In some cases, however, a combination of different binding principles is more successful than the use of a single method. Figure 27 shows an example in which adsorption, covalent binding and cross-linking are all involved.

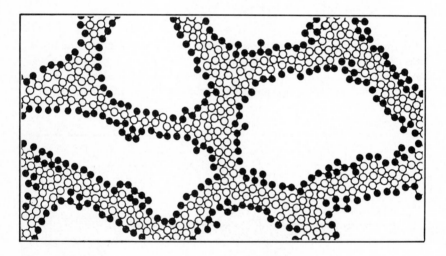

Fig. 27. Part of a porous carrier with enzymes coupled to
O = carrier subunit   ● = enzyme molecule

In the example shown (Fig. 27), the carriers are porous particles of about 50 to 150 μm diameter. Carriers of this kind can be produced, for example, by cross-linking inactive proteins (albumin, gelatine etc.). If these high-polymer substances are treated with glutaraldehyde they can then, via their amino groups, be coupled with enzymes. Thus, the pores of the carrier are lined with enzyme molecules which are to some extent cross-linked to one another and also partly bound by covalent bonds to the carrier subunits. In this way detachment of the enzyme layer lining the pores is prevented.

Entrapment   Following Prepolymer Formation

In the production of polymers the usually highly reactive monomers and the harsh reaction conditions required have a strongly deactivating effect on the biocatalysts. Above all, live whole cells may not necessarily be able to retain their viability during the process of **entrapment**   with synthetic polymers (cf. Sect. 2.5).

The use of a prepolymerization method can at least greatly reduce the danger of inactivation of the biocatalysts. The principle of this method, shown in Fig. 28, is that, as far as possible, reactions potentially detrimental to the enzymes and cells are carried out separately. The biocatalysts are added when the prepolymers have already formed but are still in a soluble state. Contact between the prepolymer and biocatalysts is allowed only after removal of the highly toxic monomers. The final polymerization leading to an insoluble matrix can then be limited to a relatively few and mild coupling reactions.

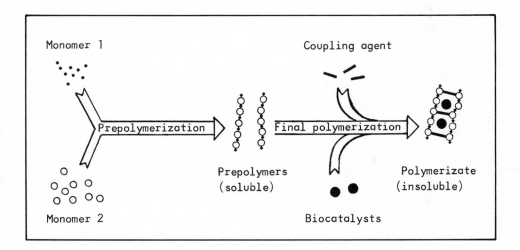

Fig. 28. Entrapment of biocatalysts after prepolymerization

Sections 9.4 and 9.5 can be referred to for further combined methods of immobilization.

2.8 Immobilization of Coenzymes

Coenzymes that are relatively tightly bound to their apoenzyme, as for example FAD to glucose oxidase, are fairly simple to immobilize since they are automatically coupled along with the apoenzyme. On the other hand, freely dissociating coenzymes such as NAD, NADP, ADP or ATP are a very different matter. They leave the apoenzyme on which, for example, they have been reduced from NAD to NADH, and move to a second apoenzyme on which they are reoxidized (regenerated).

Immobilization of only one of the apoenzymes of a freely dissociating coenzyme is of little use unless the coenzyme itself and the apoenzyme required for regeneration are also immobilized. At the same time, the coenzyme must remain sufficiently mobile to move to and from between the active centers of both apoenzymes. Attempts to solve this problem have been going on ever since the end of the 1960's; systems of industrial use are the result of recent developments.

Coenzyme Binding to Cross-linked Enzymes

In this method, the coenzyme, as shown in Fig. 29, is covalently bound to a relatively long spacer arm (e.g., 6-aminohexylcarbamomethyl-) which, in turn, is coupled to one of the enzyme molecules that are joined together by cross-linkages.

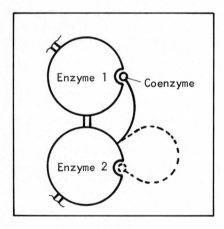

Fig. 29. Cross-linked enzymes with co-enzyme coupled via a spacer arm

For the system shown in Fig. 29, consisting of two different enzymes and a coenzyme, to function, it is essential that the coenzyme should be bound closely enough to the two apoenzymes in order to be able to reach their active centers. The statistical probability that apoenzymes with their active sites and coenzymes are arranged as shown in Fig. 29 is in fact very slight: the activities achieved in the production of such systems are therefore low.

Coenzyme Binding Next to Enzymes on a Carrier

Instead of binding the coenzyme to a cross-linked enzyme system it can be bound via a suitable spacer to a carrier. The apoenzymes must be coupled in the immediate vicinity of the coenzyme, and in such a way that their active centers, as shown in Fig. 30, can be reached by the cofactor.

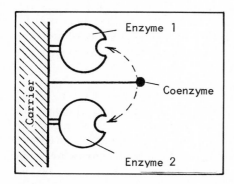

Fig. 30. Carrier-fixed system comprising two apoenzymes and a coenzyme

To make possible the coupling of the coenzyme to the carrier it must first be modified by coupling with a sufficiently long spacer. A usual method to convert NAD first to carboxymethyl-NAD and then, using carbodiimide, to 8-(6-aminohexyl)-carbamomethyl-NAD, which has the following structure:

$$H_2N-(CH_2)_6-NH-CO-CH_2-NH-\widehat{NAD}$$

This NAD-bound spacer can then be coupled via its free amino group with suitable activated carriers such as Sepharose, cellulose and others after activation with BrCN (cf. also Sect. 2.3, p. 29). Although the basic activity of the carrier-bound coenzyme can largely be retained in reactions involving soluble enzymes, this is not the case if the enzymes are also bound. The ideal configuration shown in Fig. 30, in which the active sites of both apoenzymes are readily accessible, is seldom achieved.

Molecular Enlargement of Coenzymes

So far, the most sucessful concept in the field of coenzyme immobilization has proved to be that of increasing the molecular weight and confining the enlarged molecule within a membrane. Figure 31 shows such a system. Although the coenzymes remain soluble after the procedure of enlargement they can be retained in a defined space by commercially available ultra-filters. In this form, coenzymes were used in industry for the first time, to produce L-amino acids (cf. Sect. 5.3). They could also, in principle, be used in microcapsules or liposomes, but the high sensitivity of these two forms makes the prospect of their large-scale use doubtful.

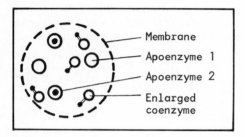

Fig. 31. System with enlarged coenzyme molecules

The molecular weight of coenzymes has been increased successfully by using polyethylene glycol (PEG), polylysine, polyethylene imine and dextran, to name only a few: in industrial applications soluble PEG with a molecular weight of 20.000 is chiefly used. The subsequent coupling is carried out as described on p. 49, using carbodiimide, following cyanogen bromide activation of the molecule.

3 Characteristics of Immobilized Biocatalysts

3.1 Activity as a Function of Temperature

The influence of temperature on the activity of native and immobilized biocatalysts is usually represented in the form of so-called optimum curves (see Fig. 32), in which activity is plotted against temperature. Often, the activity is not given in international units but as the relative activity, i.e., the quotient of the actual activity and the highest activity measured.

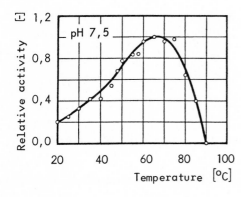

Fig. 32. Papain activity as a function of temperature.
10 min duration of trial

Unfortunately the information contained in optimum curves, such as Fig. 32, is often accepted unquestioningly and the values are taken as absolute. As a consequence, a temperature optimum taken from one of these curves is often mistakenly regarded as a constant for the biocatalyst concerned. It must always be borne in mind that this "optimum" differs according to the length of time for which the catalyst is exposed to a particular temperature (duration of measurement). The shorter the time of exposure the higher is the temperature optimum of the enzyme reaction. The optimum curves are the net result of increasing activity with rising temperature (acceleration of reaction) and a concurrently increasing inactivation (destruction of activity).

Immobilization affects both activation and inactivation of enzymes and the more complex biocatalysts. The optimum curves are in many cases altered by immobilization, depending on the method selected. At present, there is no reliable way of predicting the way in which a particular method of immobilization will affect an enzyme. For the time being, the only means of obtaining exact data on temperature relationships is practical experimentation.

The Arrhenius Diagram

The activating effect of temperature on an enzyme-catalyzed reaction can be analyzed with the help of an Arrhenius plot, in which the activity, according to convention, is plotted logarithmically against the reciprocal of the absolute temperature.

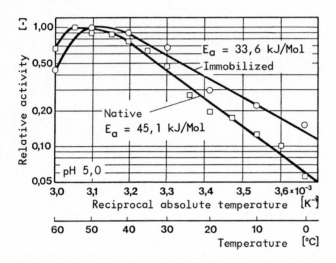

Fig. 33. ARRHENIUS plot of temperature and activity of native and immobilized β-galactosidase

It is known that high temperatures, i.e., in the left-hand portion of the curve in Fig. 33, have a highly inactivating effect. Above a certain limit - extreme left - a further rise in temperature results in a drop in activity. Conversely, at low temperatures - right-hand side of Fig. 33 - there appears to be no inactivation with rising temperature. From this portion of the curve (typically a straight line) the activation energy can be calculated as described below.

Determination of Activation Energy

The greater the activation energy, the more is the enzyme-catalyzed reaction accelerated by an increase in temperature. Expressed the other way round, at a low activation energy the reaction proceeds relatively fast, even if the temperature is not raised. Thus, a rise in temperature increases the speed of the reaction less if the energy of activation is low than if it is high.

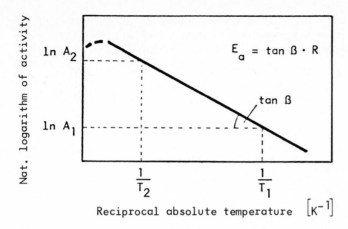

Fig. 34. Evaluation of activation energy

According to Arrhenius, the activation energy, plotted logarithmically as in Fig. 34, is directly proportional to the slope, tan ß, and to the general gas constant R.

$$E_a = tan\beta \cdot R \qquad (1)$$

From a diagram of the type shown in Fig. 34 tan ß can be expressed as the quotient of $ln A_2 - ln A_1$ and $1/T_1 - 1/T_2$. By substituting this quotient for tan ß in Eq. (1) the energy of activation can be expressed as follows:

$$E_a = \frac{ln A_2 - ln A_1}{\frac{1}{T_1} - \frac{1}{T_2}} \cdot R \qquad (2)$$

If, as is usual, the decadic instead of the natural logarithm is employed, a multiplication factor of 2.3 has to be introduced. The value for the gas constant $R = 8.3 \ J/Mol \cdot K$ can also be inserted, so that from Eq. (2) E_a can be calculated as

$$E_a = 19,1 \cdot \frac{lg A_2 - lg A_1}{\frac{1}{T_1} - \frac{1}{T_2}} \quad [J/Mol] \qquad (3)$$

Equation (3) can be rearranged as desired, for example as

$$E_a = 19,1 \cdot \frac{T_1 \cdot T_2}{T_2 - T_1} \cdot lg \frac{A_2}{A_1} \quad [J/Mol] \qquad (4)$$

The activation energy of industrial enzymes is usually in the order of 30 to 100 kJ/Mol. There are many cases on record in which the activation energy, as in the example shown in Fig. 33, was either raised or lowered by immobilization.

3.2 Stability as a Function of Temperature

On account of their protein character, biocatalysts are labile and more or less rapidly inactivated by heat. The curves in Fig. 35 show three frequently observed ways in which the activity of immobilized enzymes decreases with time at a given and constant temperature.

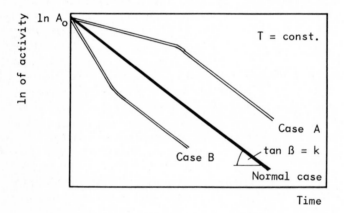

Fig. 35. Different forms of activity decrease as a function of time

The logarithmic decrease of activity with time can be considered as normal (cf. Fig. 35). This behavior is very common for single enzymes. More complex biocatalysts consisting of several enzymes of which one, particularly temperature-labile, enzyme governs the inactivation of the whole can also show this type of behavior.

Two deviations from the normal course of inactivation are shown in Fig. 35 as A and B. In each case, one of the causes may be the presence of isoenzymes of differing stability. Case A, characterized by a gradual increase in the rate of inactivation, can also be caused by an increasing microbial contamination or proteolytic breakdown with time. Case B, in which activity at first decreases quickly and then more and more slowly, can result from leaching out of unbound enzymes from the otherwise immobilized preparation.

Inactivation Coefficient and Half-life

The so-called "normal" inactivation of biocatalysts is easy to calculate, as can be seen from Fig. 35. It obeys the general logarithmic function

$$A_t = A_o \cdot e^{-k \cdot t} \tag{5}$$

in which the inactivation coefficient k is equal to the slope of the line in a plot such as that shown in Fig. 35. The coefficient k is by no means a constant, but varies according to the temperature (cf. Fig. 36).

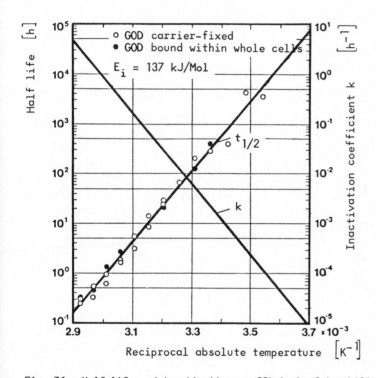

Fig. 36. Half-life and inactivation coefficient of immobilized glucose oxidase under resting conditions

The higher the temperature the greater is the value for k. Another important characteristic of the stability of a biocatalyst - and one that is easily demonstrated - is the half-life, by which we mean the time required for the activity to drop to half of its initial value under a given set of conditions. If the activity decreases according to Eq. (5) there is an inversely proportional relationship between half-life and inactivation coefficient:

$$t_{1/2} = \frac{\ln 0,5}{-k} = \frac{0,693}{k}$$ (6)

If the inactivation coefficient and the half-life are plotted semi-logarithmically against the reciprocal value of the absolute temperature, the typical result is that the two curves intersect as shown in Fig. 36.

Inactivation Energy

By using a procedure analogous to that employed for determining the activation energy (see Sect. 3.1, pp. 52f), the inactivation energy E_i can be calculated from the inactivation coefficient obtained at different temperatures. If the activity values in Eq. (4) are replaced by inactivation coefficients the equation can then be rewritten as

$$E_i = 19,1 \cdot \frac{T_1 \cdot T_2}{T_2 - T_1} \cdot \lg \frac{k_2}{k_1} \quad [J/Mol]$$ (7)

The inactivation energy of enzymes in industrial use, and of micro-organisms, is usually between 200 and 400 kJ/Mol, i.e. it is distinctly higher than the activation energy, which is generally below 100 kJ/Mol (cf. Sect. 3.1, p. 54). The difference in magnitude between the activation energy and the inactivation energy is the reason for their being predominantly activated at lower temperatures and mainly inactivated at higher temperatures. In an Arrhenius diagram (see Fig. 37) this is clearly seen in the different slope of the curves. At low temperatures, i.e., in the right-hand side of Fig. 37, activation predominates; the lower the temperature, the less important is the role of inactivation.

The stability of a biocatalyst expressed, for example, as half-life, is often greater in the immobilized form than in the native form. This is understandable because the conformation of the enzyme protein is usually stabilized by the immobilization procedure. It is not unusual, as in the immobilization of enzymes with the aid of cross-linking reagents for instance, that intramolecular bonds are established between different parts of the protein chain in addition to the intermolecular bonds between the enzyme molecules. In most cases this serves to reduce the danger of chain rupture due to thermal oscillations. However, immobilization procedures weaken the conformation of the enzyme chain in some cases, for example if the newly formed bonds create greater tension at weak points in the molecule.

The insufficiency of our knowledge of the situation in the molecule combined with inability to govern the exact site via which the biocatalyst is bound, means that influencing stability is still largely a matter of chance.

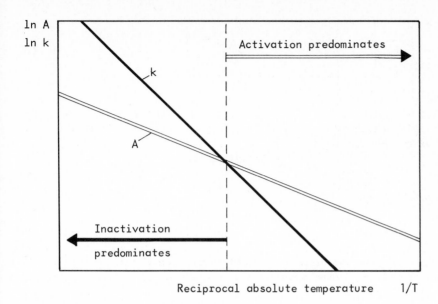

ln A
ln k

Activation predominates

k

A

Inactivation

predominates

Reciprocal absolute temperature   1/T

Fig. 37. Activity and inactivation coefficient as a function of the reciprocal absolute temperature

3.3 Temperature Optimum in Long-term Processes

In the two preceding Sects. 3.1 and 3.2 we discussed the activating and inactivating effects of temperature on the activity of biocatalysts. It was shown that these two effects act in opposite directions and have to be taken into consideration when determining the optimum reaction temperature.

Figure 38 illustrates, again (in this case for an immobilized gluco-amylase) the well-known fact that the enzyme has a higher initial activity at 50° C than at 40° C. At the lower temperature, however, its activity naturally drops very much more slowly because the enzyme is more stable at this temperature.

If the activity of an enzyme is known, and if it remains constant, the quantity of substrate converted in a given period of time can easily be calculated. Activity is the quantity of substrate converted per time; if it is multiplied by time, the total amount of substrate converted within this period is obtained. If activity is plotted against time, the quantity of substrate converted is given by the area below the activity curve. This also applies if, as shown in Fig. 38, the activity changes with time.

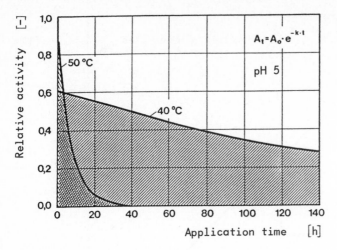

Fig. 38. Activity of an immobilized glucoamylase as a function of application time with temperature as parameter.

The quantity of substrate turnover, shown as the hatched area below the curve in Fig. 38, is distinctly larger for a very short period of time (e.g., 2 h) at 50° C than at 40° C. After longer use (e.g., 120 h), however, the total amount of substrate broken down by the glucoamylase is much larger at 40° C than at 50° C. Figure 38 shows only relative values for activity and therefore the areas beneath the curves indicate only relative amounts of substrate converted or products formed.

Provided that the decline in activity obeys, at least approximately, Eq. (5), the areas beneath the curves in Fig. 38 can be calculated by integrating Eq. (5) to give the quantity of substrate turnover T within the time period between t_1 and t_2.

$$T = \int_{t_1}^{t_2} A_o \cdot e^{-k \cdot t}\, dt \qquad (8)$$

Following integration this becomes

$$T = A_o \cdot \frac{1}{-k} \cdot e^{-k \cdot t} \Big|_{t_1}^{t_2} \qquad (9)$$

The reaction (= use) of the enzyme normally begins at the time $t = 0$. t_2 is then equal to the length of use and can simply be termed t. Substituting $t_1 = 0$ and $t_2 = t$, the amount of substrate turnover is given by Eq. (10).

$$T = \frac{A_o}{k} (1 - e^{-k \cdot t}) \tag{10}$$

With the help of Eqs. (9) and (10) and the experimentally determined inactivation coefficient it is possible to calculate the expected substrate turnover for any desired temperature and length of use.

The optimum reaction temperature for the long-term application of immobilized biocatalysts generally lies below the value shown in diagrams such as Fig. 32 (see p. 51), or given as the "optimum temperature" in many of the descriptions of enzymes.

3.4 The Influence of pH Value

The pH value has a considerable influence on the efficiency and stability of all enzymes. A distinction can be made between direct effects on the charged state of the substrate, effectors and functional groups of the active center, and indirect effects on groups in the vicinity of the active center or on the buffer employed.

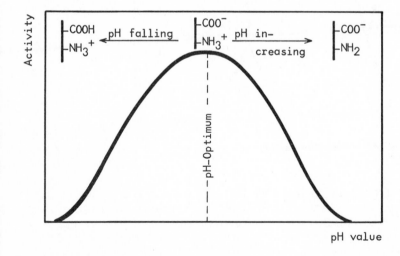

Fig. 39. Typical curve of enzyme activity vs pH value. Amino and carboxyl groups at different pH values

Figure 39 shows a typical pH-activity curve and the usual alterations in the charge on an NH_2 group and a COOH group caused by raising and lowering the pH. In this example the overall state of equilibrium between negative and positive charges is also the most active (pH optimum), although this is by no means always so. The ionic conditions under which an enzyme can exert its optimum effect depend on the enzyme in question. On account of the large number and different types of charged groups involved, the situation for enzymes is, in fact, far more complicated than in this model case.

The position of the pH optimum of an enzyme is by no means identical under all conditions. Among other factors causing slight shifts are temperature and buffer.

Since the procedure of immobilization usually has a variety of effects on the conformation as well as on the state of ionization and dissociation of an enzyme and its environment, it is not uncommon for it to result in changes in the relationship between pH and stability and activity. The same holds true for complex biocatalysts containing more than one enzyme, such as living cells. For the sake of simplification we shall confine ourselves here to a consideration of single enzymes.

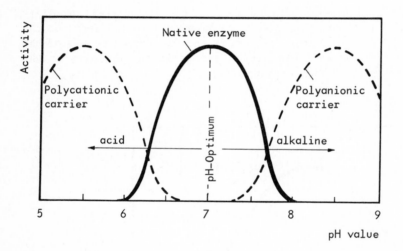

Fig. 40. Influence of charged carriers on the pH dependence of bound enzymes

We are still far from being able to predict or to calculate in advance the influence of the various methods of immobilization on the relationship between pH and activity of an enzyme. Nevertheless, in a few exceptions the general trend can be foreseen. It is known, for example, that if an enzyme is coupled to a polyanionic carrier the pH optimum usually shifts in the alkaline direction whereas if the carrier is polycationic the shift is in the acid direction. Figure 40 shows this commonly observed effect of carrier charge on the pH-activity curve.

The charge on the carrier surface changes the microenvironment of the enzyme bound to it. A polycationic (positive) charge on the carrier, as illustrated in Fig. 41, has the effect that more ions with an opposite (i.e., negative) charge are drawn into the vicinity of the bound enzyme. An enzyme protein that develops its maximum activity at pH 7, for example, will encounter this pH value in its microenvironment when the macroenvironment has a higher concentration of H^+ ions, or a lower pH than 7. Thus, using a polycationic carrier the pH optimum shifts toward a lower pH in the bulk of the reaction solution where the pH is generally measured. A shift of this kind can amount to as much as two pH units.

The "normal" case described, and illustrated in Figs. 40 and 41, can be modified by a large variety of factors. For example, the insertion of a spacer between the carrier and the enzyme by increasing the distance between them may mean that the enzyme protein enters a completely different microenvironment, with the result that the pH dependence of the activity of the enzyme is altered. Further, the position of the active center on the enzyme protein, the size of the enzmye molecule and the site at which the enzyme is bound are factors involved in determining the distance of the catalytic center from the carrier surface, and thus whether or not it will be affected by the surface charges.

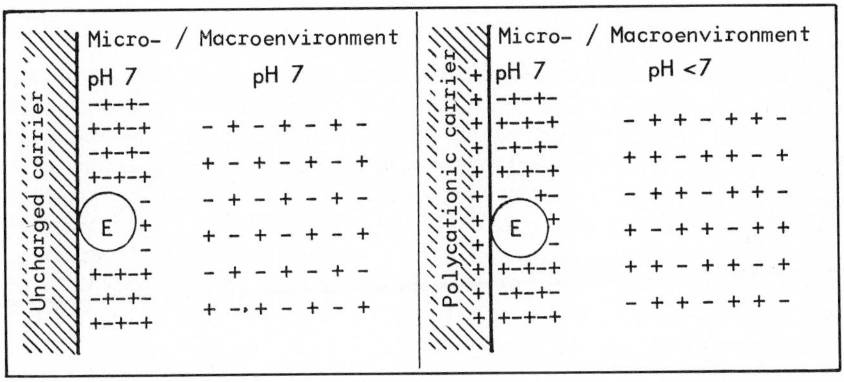

Fig. 41. Ions in the vicinity of an enzyme bound to an uncharged and to a polycationic carrier

Not only the activity of biocatalysts but their stability as well is influenced by the pH value. Curves depicting the relationship between stability and pH often closely resemble the pH-activity curves (cf. Fig. 39). However, the pH optimum for maintaining maximum stability may differ considerably from that for maximum activity. The stability optimum is usually much broader than the activity optimum.

Stability to pH changes caused by acids or alkalis is as a rule improved by immobilization, although it may occasionally be lowered. Here again, far too little is known about the exact nature of events in the enzyme protein and about the vast number of possible interactions for accurate predictions or calculations to be made in advance.

3.5 The Influence of Substrate Concentration

The velocity of an enzyme reaction is decisively influenced by the concentration of its substrate. In most cases this relationship obeys Michaelis-Menten kinetics. The typical shape of a curve for an enzyme with this type of kinetics is shown in Fig. 42 as the continuous black line. This so-called substrate-saturation curve is obtained by plotting the velocity of the reaction against substrate concentration. At high substrate concentrations the reaction practically obeys zero-order kinetics, i.e., it proceeds with an almost constant velocity (V_{max}), independently of the substrate concentration. At very low substrate concentrations the speed of reaction is approximately proportional to the substrate concentration, corresponding to first-order kinetics. Between these two extremes there is a reaction with mixed-order kinetics. The substrate-saturation curve obeys the equation

$$v = \frac{V_{max} \cdot S}{K_m + S} \qquad (11)$$

It should be emphasized that the velocity of the reaction, v, is not a velocity in the physical sense (distance per time), but a turnover rate (amount per time). K_m is the Michaelis constant, defined as the dissociation constant of the enzyme-substrate complex; it has the dimensions of a concentration (e.g., mMol/l) and corresponds to the substrate concentration at which half the maximum velocity is achieved. At this concentration the enzyme is half-saturated with substrate, i.e., statistically, 50 % of the active centers are, at this point, bound to the substrate to give an enzyme-substrate complex. A low K_m value thus indicates a high affinity of the enzyme for its substrate.

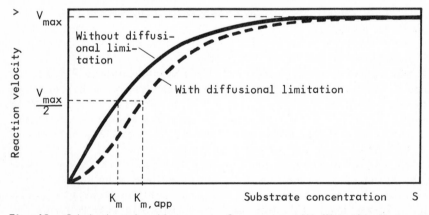

Fig. 42. Substrate saturation curve of an enzyme with Michaelis-Menten kinetics and an enzyme with limited diffusion

If access of the substrate to the active center of the enzyme is obstructed by diffusion barriers, deviations from Michaelis-Menten kinetics are observed (cf. Fig. 42). The velocity of the reaction no longer obeys Eq. (11) but a much more complicated function. In spite of this, a Michaelis constant is often given for reactions in which diffusional limitations are involved, either because the deviation from Michaelis-Menten behavior has been overlooked, or because it is only slight. It is usual to term Michaelis constants of immobilized enzymes "apparent." In this way, it is made clear that this is the K_m actually measured, without entering into a discussion as to whether it may be different from the value measured when the enzyme is in its native state.

In fact, the immobilization of an enzyme often brings about a rise in the apparent Michaelis constant, although the reverse case, i.e., a drop in the value of the constant, is also not uncommon. Table 17 gives some K_m values from the literature for the enzymes glucose oxidase and ß-galactosidase (lactase).

An increase in the affinity of an enzyme for its substrate (= a drop in the K_m value) is to be expected if the enzyme and the carrier material are oppositely charged. A lowering of the K_m value can also result from changes in conformation that improve access of the substrate to the active center of the enzyme.

Table 17. Literature data on Michaelis constants of glucose oxidase and ß-galactosidase in the native and the carrier-fixed form

Enzyme	Carrier	K_m values native	immobil.	Reference
Glucose oxidase	Sepharose	19 mM	20 mM	D´Angiuro and Cremonesi (1982)
Glucose oxidase	Membrane	48 mM	2 mM	Tsuchida and Yoda (1981)
Glucose oxidase	Gelatine	20 mM	25 mM	Hartmeier and Tegge (1979)
Glucose oxidase	Mycelium	20 mM	25 mM	Döppner and Hartmeier (1984)
Glucose oxidase	Ceramics	34 mM	16 mM	Richter and Heinecker (1979)
ß-Galactosidase	Whole cells	32 mM	59 mM	Decleire et al. (1985)
ß-Galactosidase	Glass	16 mM	19 mM	Greenberg and Mahoney (1981)
ß-Galactosidase	Sepharose	112 mM	120 mM	Friend and Shaghani (1982)
ß-Galactosidase	Protein	19 mM	20 mM	Hartmeier (1977)
ß-Galactosidase	Polyacrylamide	46 mM	46 mM	Kobayashi et al. (1975)
ß-Galactosidase	Alginate	102 mM	119 mM	Banerjee et al. (1984)

A very popular way of representing enzyme kinetics is to plot the reciprocal of the reaction velocity (1/v) against the reciprocal of the substrate concentration (1/S). This double reciprocal plot, shown in Fig. 43, has the advantage that if an enzyme reaction obeys Michaelis-Menten kinetics it can be represented by a straight line according to the following equation

$$\frac{1}{v} = \frac{K_m}{V_{max}} \cdot \frac{1}{S} + \frac{1}{V_{max}}$$ (12)

The characteristic values of enzyme reactions, V_{max} and K_m, can be read off as reciprocal values from the LINEWEAVER-BURK diagram as shown in Fig. 43. A reaction with restricted diffusion, especially at lower substrate concentrations (to the right in Fig. 43), deviates appreciably from the unrestricted reaction. The higher the substrate concentration, the less is the velocity of the reaction reduced by a drop in concentration resulting from diffusional limitations.

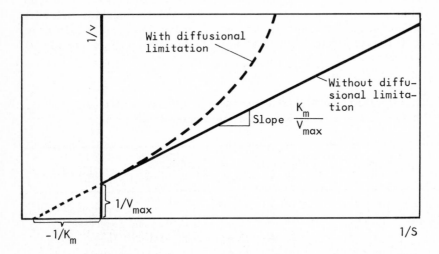

Fig. 43.   LINEWEAVER-BURK plot of an enzyme reaction with and without limited diffusion

A disadvantage of the LINEWEAVER-BURK method is the accumulation of measurements toward the abscissa. In addition, exaggerated importance is attached to the measurements obtained at lower substrate concentrations and consequently also to the errors attaching to them. The less commonly used linear representations according to EADIE and HOFSTEE, or to HANES, are often better suited for the determination of enzyme kinetic data than the LINEWEAVER-BURK diagram. Further details of these alternative methods can be found in the relevant literature on enzyme kinetics.

3.6 The Influence of Diffusion

The influence of diffusion on enzyme-catalyzed reactions has already been touched upon in the preceding section (Sect. 3.5). We shall now take a closer look at this effect and see how it can be quantified. First of all, a distinction has to be made between diffusional limitations to external and to internal mass transfer. Barriers to external diffusion build up when there is insufficient movement in the medium surrounding the biocatalysts, and may, for example, take the form of a film of static fluid. External mass transfer will be dealt with in connection with reactors in Chap. 4. Internal diffusion barriers, which are our main concern at this point, are chiefly caused by the enveloping matrices of the biocatalysts.

Fick's Law

The velocity of diffusion, v_d, of an enzyme substrate obeys Fick's first law:

$$v_d = D_e \cdot \frac{F}{r} \cdot \Delta S \qquad (13)$$

Here, too, as with the velocity of an enzyme reaction, the diffusion velocity is not a velocity in the physical sense (distance per time), but is a rate expressing the quantity of mass transfer per unit of time, i.e., in mMol/s. D_e represents the effective diffusion constant in cm^2/s, F the diffusion surface area in cm^2, r the diffusion distance in cm and S the difference between substrate concentration at the beginning and the end of the diffusion distance in $mMol/cm^3$: as a rule this is the difference between the substrate concentration outside the matrix, S_{ex}, and at the site of the enzyme, S_{en}. Equation (13) can thus be rewritten as

$$v_d = D_e \cdot \frac{F}{r} \cdot (S_{ex} - S_{en}) \qquad (14)$$

In a given system with unchanging values for D_e, F and r, these three quantities can be expressed collectively as the permeability factor P (= $D_e \cdot F/r$), which is given the dimension cm^3/s. Equation (14) can thus be simplified to

$$v_d = P \cdot (S_{ex} - S_{en}) \qquad (15)$$

S_{en} is decisive for the enzyme reaction and must be substituted in Eq. (11) to calculate the speed of the diffusion-inhibited reaction. Since S_{en} cannot be measured directly, it has to be calculated from measurable or known values for S_{ex}, D_e, F, r and the diffusion velocity. The diffusion velocity v_d can be measured because it is identical with the velocity of the enzyme reaction (enzyme activity). When the reaction is in a steady state the quantity of substrate diffusing and the quantity converted must be identical. For an enzyme reaction hindered by diffusional limitations the following rather cumbersome equation results; it describes the curve shown in Fig. 42 (see p. 62).

$$v = \frac{V_{max} \cdot \left(S_{ex} - K_m - \dfrac{V_{max}}{P} + \sqrt{\left(K_m + \dfrac{V_{max}}{P} - S_{ex}\right)^2 + 4 K_m \cdot S_{ex}} \right)}{K_m + S_{ex} - \dfrac{V_{max}}{P} + \sqrt{\left(K_m + \dfrac{V_{max}}{P} - S_{ex}\right)^2 + 4 K_m \cdot S_{ex}}} \qquad (16)$$

The situation can be further complicated, for example, by differences in the distance of the enzyme from the matrix surface, or by nonhomogeneous distribution of the enzymes or more complex biocatalysts within a matrix.

The Thiele Modulus

The Thiele modulus φ provides us with a reliable, dimensionless index defining the influences of diffusion. For a first-order reaction, in which diffusion takes place across a flat surface, the Thiele modulus is defined as

$$\varphi = r \cdot \sqrt{\frac{k}{D_e}} \qquad (17)$$

where r is the diffusion distance in cm, k the velocity constant for the reaction in s^{-1}, and D_e the effective diffusion constant in cm^2/s, already mentioned above.

For Michaelis-Menten kinetics, normally obeyed by enzyme reactions, the quotient of V_{max} and K_m can be introduced instead of the velocity constant k. The Thiele modulus can then be rewritten as

$$\varphi = r \cdot \sqrt{\frac{V_{max}}{K_m \cdot D_e}} \qquad (18)$$

However, with immobilized biocatalysts diffusion is often into a spherical structure and not across a flat surface. In this case, one-third of the radius of the sphere is substituted in the Thiele modulus

$$\varphi = \frac{r}{3} \cdot \sqrt{\frac{V_{max}}{K_m \cdot D_e}} \qquad (19)$$

If an enzyme reaction obeys Michaelis-Menten kinetics, the effectiveness of an immobilized biocatalyst can be determined by purely mathematical means as a function of the Thiele modulus, with the effectiveness η being defined as the ratio of the reaction velocity of the immobilized biocatalyst to that of the same biocatalyst in its native state.

Figure 44 shows the calculated relationship between effectiveness and Thiele modulus, with substrate concentration as parameter. This relationship applies in each case to a given ratio of total mass transfer to diffusion, which, in turn, is characterized by the dimensionless Sherwood number Sh. It would exceed the scope of an introductory text to go into this problem in more detail. In principle, the effectiveness is reduced less as the Sherwood number increases or, in other words, the less the mass transfer depends on diffusion. Figure 44 describes the situation for an average Sherwood number of 8; the figure also contains a few measurements obtained with glucose oxidase (GOD), mycelium-bound and entrapped in alginate beads.

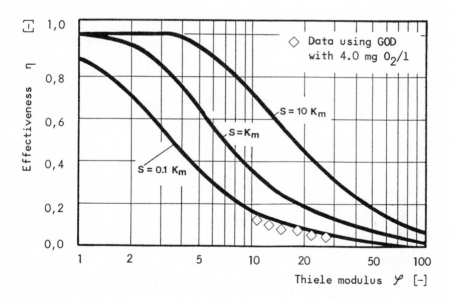

Fig. 44. Effectiveness as a function of the Thiele Modulus

The effectiveness of a process catalyzed by immobilized biocatalysts increases (as is clearly shown in Fig. 44) with the substrate concentration. The higher the ratio of substrate concentration to K_m, the less is the reaction velocity reduced by the drop in concentration (due to obstructed diffusion) from matrix surface to enzyme site.

3.7 Other Physical Properties

In addition to data on enzyme kinetics, information regarding the mechanical strength of immobilized biocatalysts is very important in connection with their practical application.

Breaking Strength

The group of J. Klein (Braunschweig) has developed a setup (shown diagrammatically in Fig. 45) for measuring the breaking strength of biocatalysts encapsulated in polymers. The sphere is placed on a rigid surface and compressed by a slowly descending pressure plate. The force required is measured and recorded. At the breaking point of the sphere the pressure force suddenly drops due to release of the opposing pressure. But as the plate continues to move downward pressure gradually builds up again due to resistance from the fragments of the broken bead. Figure 46 is a recording from such a test.

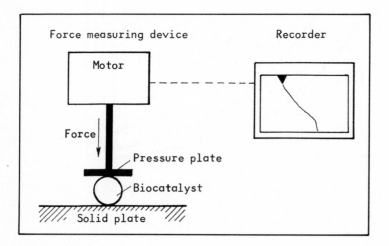

Fig. 45. Device for measuring the mechanical stability of biocatalyst beads

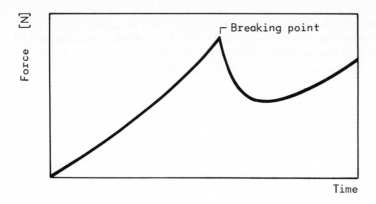

Fig. 46. Recording of an experiment according to Fig. 45

Pressure Buildup in a Packed Bed

One of the most popular ways of using immobilized biocatalysts is to pack them in a column as a so-called "packed bed". The pressure resistance that has to be overcome during flow through the column, as Fig. 47 shows, either rises in direct proportion to the height of the column, or it may rise disproportionately. The first situation arises if the biocatalyst particles are rigid and nondeformable, whereas if the particles are compressible the pressure rises disproportionately with increasing column height. From a certain height onward, which to some extent depends on the size and nature of the compressible particles, the packed bed can no longer be used.

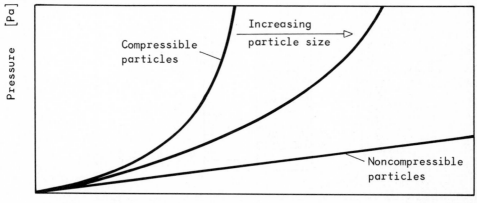

Fig. 47. Buildup of pressure as a function of layer depth using compressible and noncompressible biocatalyst particles

In the case of noncompressible particles, under strictly standardized conditions, the pressure buildup (indicated by the difference in pressure between inflow and outflow of the packed bed) can be described by a simple function. The pressure buildup then conforms to the equation

$$P = k_p \cdot h \tag{20}$$

in which h is the height of the packed bed (given in m for example), k_p is the coefficient of pressure buildup or pressure drop (with the dimension, for example, of Pa/m). The coefficient k_p determines the slope of the straight line in Fig. 47: it is not a constant, but apart from depending on the nature of the biocatalysts particles it is influenced by the surface speed v_o (given in m/s) and the viscosity u (in m^2/s) of the solution being forced through the bed. Preliminary efforts have been made to express these factors in mathematical terms, but practical experimentation is more or less indispensable for each particular case in which compressible immobilized biocatalysts are used in practice.

In a setup like the one depicted in Fig. 48, the pressure increase can be measured for varying heights of packed bed, flow rate, temperature and so on. With the help of the data obtained, optimized packed beds can be constructed for industrial application.

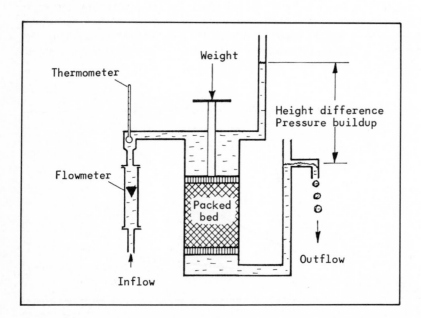

Fig. 48. Device for testing the pressure build-up caused by biocatalyst particles in a packed bed

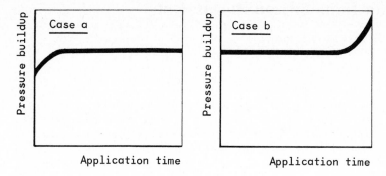

Fig. 49. Typical changes in pressure buildup as a function of application time

During prolonged use of immobilized biocatalysts in a packed bed, changes in the pressure buildup are frequently experienced. Figure 49 shows two commonly encountered examples: normal ageing brings with it a gradual increase in pressure, and sometimes a particularly rapid buildup in the first phase (see Fig. 49, case a). Case b of Fig. 49 may be caused by obstruction of the packed bed by microorganisms or by blockage with substrate particles.

6 Reactors for Immobilized Biocatalysts

The effect of diffusional limitations on external mass transfer in connection with the use of immobilized biocatalysts has already been referred to in Sect. 3.6. External mass transfer is essential for the transport of the dissolved substances (substrate) to the immobilized biocatalysts. The further transport within the biocatalysts to the reaction site, which we call internal mass transport, cannot be directly influenced by the nature of the bioreactor.

The main task of the reactor is, by ensuring adequate relative movement between the biocatalysts and their surrounding medium, to increase external mass transfer to such a degree that the external diffusional limitations play practically no role (see Fig. 50). The thickness of the stagnant fluid layer surrounding the biocatalysts is the chief cause of resistance to external transport and can be reduced by increasing the stirring or flow intensity. As Fig. 50 shows, an improvement in fluid movement leads to a decrease in the apparent Michaelis constant and a rise in the rate of the reaction.

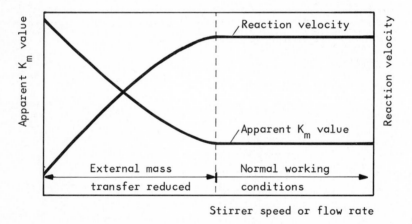

Stirrer speed or flow rate

Fig. 50. Apparent K_m value and reaction velocity as functions of stirrer speed or flow rate

In addition to creating optimum hydrodynamic conditions the reactor has to provide for retention of the immobilized biocatalysts. These and often other functions, such as supplying oxygen or removing carbon dioxide, have to be carried out under conditions that offer maximum protection to the, for the most part, sensitive biocatalysts.

4.1 Stirred Reactors

The stirred reactor is the type most widely employed in fermentation techniques. A high stirring efficiency ensures rapid mixing and highly efficient oxygen transfer in the case of aerobic fermentation. Stirred tank reactors are not commonly used in combination with immobilized catalysts since intensive stirring such as is required in the aerobic processes of classical fermentation is seldom necessary. The considerable shearing forces frequently developing in stirred reactors can be a source of wear and abrasion to the biocatalyst particles. A disadvantage of these reactors as compared with the packed bed type is that, because the particles have to be in suspension, only much smaller quantities of biocatalyst per volume can be used.

Advantages of the stirred tank that recommend its use with immobilized biocatalysts are its simple and cheap construction, and the fact that it has been well investigated. However, the major use for this type of reactor is for acid-, alkali-, or oxygen-consuming reactions. As a rule, thicker, viscous substrates can, as a rule, also be processed with no difficulty. In principle, the stirred tank reactor can be used either for batchwise or continuous reactions. In the batch procedure the biocatalysts are separated or filtered off at the end of the reaction for use in the next batch. In the continuous procedure, in which the addition of substrate and withdrawal of product go on without interruption, the immobilized biocatalysts must either be retained by a sieve across the outlet or continuously reentered together with fresh substrate.

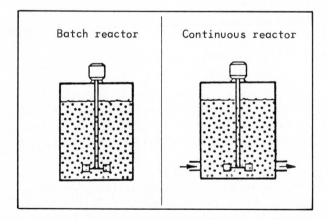

Fig. 51. Batch and continuous stirred tank reactors with immobilized biocatalysts

In using immobilized biocatalysts the aim is often to establish a continuous process in which a steady state prevails, i.e., a state in which the actual concentrations of substrate and product in the reactor remain the same at all times (see Fig. 52).

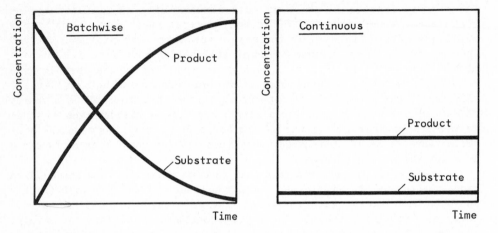

Fig. 52. Substrate and product concentrations as functions of time in batch-wise and continuous fermentation

For the homogeneous, one-step reaction (Fig. 51) the rate of inflow, f_1, equals the rate of outflow, f_2, so that no distinction need be made between the two values and they are referred to as the flow rate, f, in 1 per h (or m^3/h). The ratio of the flow rate to the working volume of the reactor is termed the specific flow rate or the dilution rate, D.

$$D = \frac{f}{V} \quad [h^{-1}]$$

(21)

The dilution rate, D, is given the dimension h^{-1}, and is sometimes known as the space velocity, especially in connection with bed reactors. D gives an indication of how often the reactor has, theoretically, to be refilled per hour with the fluid flowing through it. The reciprocal of the dilution rate is the mean residence time, t_m, which tells us the average length of time that the inflowing fluid spends in the reactor.

A highly important criterion of efficiency in industrial processes is the volumetric productivity, P_v, which is obtained by multiplying the specific flow rate by the product concentration, A, in g/1 (or kg/m^3). In the homogeneous, continuous stirred tank discussed here, the product concentration is uniform throughout the entire contents and outflow.

$$P_v = D \cdot \bar{A} \quad [g/1 \cdot h]$$

(22)

The volumetric productivity indicates how many g or kg of product per 1 or m^3 of effective reactor volume are produced hourly.

Continuous operation need not necessarily be superior to a batchwise procedure. One disadvantage of homogeneous, continuous processes is the low actual substrate concentration and high product concentration in the reactor. If the affinity of the biocatalyst for the substrate is low (high K_m value) and if the reaction is inhibited by the product, only very small reaction velocities are achieved for the overall process. In batchwise procedures, however, the higher reaction velocities at the higher substrate concentrations prevailing at the beginning of the reaction can be exploited, while product inhibition and lower reaction velocity due to lack of substrate are confined to the end phase of the run (see Fig. 52). Microbial contamination can become a problem in continuous processes, especially at high mean residence times (low specific flow rates).

4.2 Loop Reactors

In loop reactors the contents are forced to circulate loopwise. This is usually achieved by building into the reactor tower a "conducting" cylinder which directs the flow, produced either by air, propeller stirrer or liquid jet, into a closed loop.

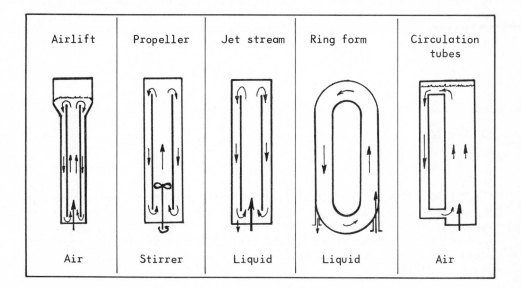

Fig. 53. Different types of loop reactors

The various names of loop reactors, i.e., airlift-, propeller- or jet-loop indicate the means by which circulation is produced. Other ways of obtaining loop circulation are to construct the reactor itself as a ring, or to add an encircling tube to the main reactor. The forms described are shown schematically in Fig. 53: all of them can be operated with reversed flow if desired.

Immobilized biocatalysts can be circulated with the substrate in loop reactors, although in some cases they are confined to one region of the flow chamber. For such purposes coarse particles are the most suitable since, they can be withheld in the desired part of the reactor by the insertion of a simple sieve.

The loop reactors shown in Fig. 53 exist in many further variations and combinations. Most loop reactors have a tower form, usually with a height to diameter ratio exceeding 5. Employed in the airlift form, loop reactors of this kind are used in the production of single cell protein and, on a large scale, for sewage treatment.

The widening of the upper portion of the airlift reactor shown in the diagram facilitates the escape of the rising, exhausted gas bubbles. The reduced flow rate in this part of the reactor resulting from the increased diameter gives the bubbles more time to escape. If too many bubbles are drawn downward at this point, circulation is hindered by the resulting upward lift. The hydrostatically driven circulation may even come to a halt if (due to the gas bubble volume) the density in the successive portions of the reactor equalizes.

An advantage of the loop reactor is that mass transfer can be achieved with only little or moderate expenditure of energy. Since laminar flow conditions prevail in reactors of this kind the biocatalysts are exposed to little shearing stress. On the other hand, research and development in the field of loop reactors are still in their early stages, which means that few reliable data are available with regard to industrial application and scaling up.

4.3 Bed Reactors

Bed reactors are very popular for carrying out reactions involving particulate biocatalysts. In most cases they are used for continuous processing. Depending on the nature of the particles constituting the bed and the way in which the substrate flows through it the reactors are termed packed bed, fluidized bed reactors, and well-mixed layer reactors (see Fig. 54).

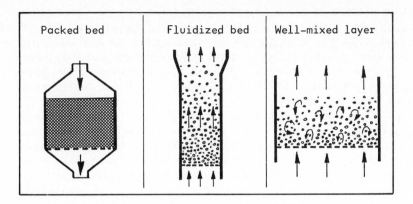

| Packed bed | Fluidized bed | Well-mixed layer |

Fig. 54. Construction of some bed reactors

Packed Bed

The packed bed reactor is the most popular of all bioreactors for immobilized biocatalysts since it permits the use of the catalysts at the highest possible density. As a consequence, relative to the available reactor volume, the highest possible substrate conversion per unit time is attainable. The volumetric productivity is therefore usually higher in the packed bed than in any other type of reactor. However, by no means all immobilized biocatalysts are suitable for use in a packed bed.

The mechanical properties of the particles that are pressed together to form the bed should be such that as little pressure as possible is required to force the substrate through it. Even after prolonged use, no channels or other irregularities should appear in the packed bed since these would impede regular flow. For gas-producing reactions such as ethanol production with yeast cells, the problem arises that the liberated CO_2 prevents optimal contact between substrate and particles.

Within a packed bed a gradient establishes itself, with a higher concentration of substrate at the inflow and a lower concentration at the exit. The distribution of residence times of the different parts of the fluid in the reactor can be held within very close limits, which makes possible an almost complete transformation of substrate into products. For reactions in the course of which the pH alters, the packed bed can only be employed if the pH-range within the bed does not exceed the limits tolerated by the biocatalysts.

Fluidized Bed

In contrast to the packed bed, the fluidized bed reactor is only loosely filled with biocatalyst particles (see Fig. 54). The substrate enters from below and is forced upward through the fluidized bed, whereby the biocatalysts are kept in a state of loose suspension by the substrate flow, given the condition that the density of the particles is greater than that of the substrate. Furthermore, retention of the biocatalysts in the fluidized bed is only possible if the medium is not too viscous and the flow rate not too high. In order to reduce the danger of flushing-out of the particles, the speed of the fluid front in the upper portion of the reactor is often reduced by a conical widening.

An important condition for efficient transformation in the fluidized bed is the plug flow of the liquid in its passage through the bed, i.e., the liquid front should move through the bed like a plug, with no back-mixing. Thus, ideally, the residence time of all substrate molecules should be equal. Maintenance of plug flow behavior is particularly difficult in scaling up the process. In gas-producing reactions, even flow through the bed is practically impossible to achieve, and in such cases the fludized bed approaches a well-mixed bed or a well-mixed layer.

Well-mixed Layer

In a well-mixed layer reactor, a large degree of backmixing is tolerated or even deliberately evoked. The formation of a well-mixed layer can be promoted by a construction with a high ratio of height to diameter, by built-in devices, by the introduction of a gas, or by reactions in which gas is liberated. Due to the broad distribution of residence times, the well-mixed layer reactor is chiefly used for the incomplete substrate conversion required in some fields of application. It is difficult to draw a line between fluidized bed with plug flow and well-mixed layer with almost ideal conditions of mixing (comparable with those achieved in stirred reactors), since practically every transitional stage between the two extremes can be encountered.

4.4 Membrane Reactors

In principle, membrane reactors are used for separating low- from high-molecular substances by means of a membrane with extremely fine pores (ultrafilter membrane). This type of reactor is particularly useful for enzyme reactions since the size of enzyme molecules often differ considerably from that of the product molecules. With the enzyme and substrate on one side of the membrane, a slight pressure (50,000 to 5 million Pa) is applied; this results in obligatory convective transport of the low-molecular product through the pores of the membrane. Commercially available ultrafilter membranes retain molecules of 500 to 300,000 MW, i.e., all sizes of enzyme molecule possible.

Biocatalysts are usually employed in membrane reactors in the water-soluble, native form. In spite of this, according to the definition given in Sect. 1.4, it is justifiable to speak of immobilized biocatalysts since the presence of the ultrafilter membrane limits the space in which the catalytic reactions take place. The biocatalysts are, so to speak, immobilized within the membrane reactor.

There are two basic forms of membrane reactor. The construction and function of the hollow fiber reactor or capillary membrane reactor have already been illustrated in Fig. 25. The second common form is the membrane tank reactor, also termed sheet membrane reactor on account of its sheet-like membrane. The construction and working principle of such reactors are shown in Fig. 55.

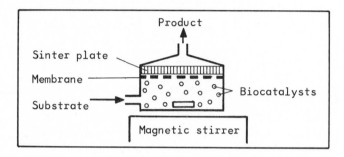

Fig. 55. Membrane tank reactor

Ultrafilter membranes as a rule consist of synthetic polymers on the basis of polyamide or polyether sulfon. Membranes on a cellulose basis are of limited applicability in enzyme reactions because many technical enzymes exhibit cellulolytic activity and would destroy the membrane. The actual fine-pore membrane layer is usually mounted on a supporting layer of microporous polymer. An asymmetrical membrane arrangement with the finest pores toward the pressure side of the reactor has proved especially successful: the flow rate is higher, the tendency to block-up is lower, and cleaning is simpler than with symmetrical arrangements.

In all membrane reactors the fluid on the pressure side must be kept in motion in order to create a high relative velocity between reaction solution and membrane. In practice this is achieved by continuous pumping through tubular membrane reactors (see Fig. 56), or by stirring the membrane tank reactor. In this way blockage of the membrane can be prevented. The forced flow ensures that the enzyme does not accumulate at the membrane to such an extent that the solubility limit is reached. The reactor is usually filled via a sterile filter in order to avoid bacterial contamination.

A considerable advantage of the membrane reactor lies in the fact that the biocatalysts can be used in their native form, thus avoiding exposure to the usual inactivating steps. In the membrane reactor, comparable to the situation in a natural cell, the biocatalysts are held back by a membrane system. In this way it becomes possible for reactions to proceed continuously or to be repeated at will.

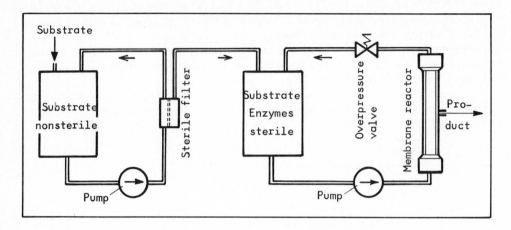

Fig. 56. Experimental setup for substrate conversion in a hollow fiber reactor

In addition to the use of native, water-soluble enzymes in membrane reactors, which is already practiced on an industrial scale, native whole cells or even carrier-bound biocatalysts can also be used in this type of reactor. A variant of the latter procedure is the use of an ultrafilter membrane to which the enzymes are coupled.

4.5 Special Forms of Reactors

Apart from the reactor types described in the preceding Sects. 4.1 to 4.4, numerous other forms of reactors for use with immobilized biocatalysts are known from the literature. In addition, combinations of the various types of construction such as a stirred tank reactor connected to a packed bed reactor are also in use. Figure 57 shows a few, widely differing examples of special reactor forms. None of the three examples illustrated has yet attained industrial importance.

The sieve-stirrer reactor sketched in Fig. 57a was suggested by Havewala and Weetall (1973). In it, carrier-bound enzymes are housed in the blades which consist of chambers surrounded by sieve-like material. When the substrate is stirred with the enzyme-filled blades, and if it can be prevented from circulating with the blades, a strong relative movement between substrate and immobilized catalysts is achieved which allows the desired reaction to proceed without external diffusional limitation.

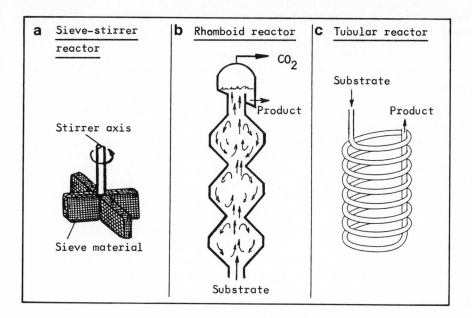

Fig. 57a–c. Some special forms of reactors

Figure 57b shows the rhomboid reactor developed by Fukushima and Yamade (1982), primarily intended for alcoholic fermentation with immobilized whole cells. By means of a vertical series of rhomboid chambers connected by narrow constrictions, circulation of the biocatalysts and substrate is achieved as shown in the diagram. Carbon dioxide produced in the fermentation process rises through the center of the rhomboid reactor and escapes at the top. The gas itself is the driving force for circulation.

A whole series of reactors has been developed on the basic plan of the tubular reactor shown in Fig. 57c, although, to date, no reactor of this type has risen to industrial importance. In one specialized form of tubular reactor the biocatalysts are bound to the inner wall of the tube. Reactors of this kind are suitable for use in analytical processes (e.g., built in autoanalyzers). Nylon or glass, to which many enzymes can be coupled relatively easily, are suitable materials for constructing the tubes.

5 Industrial Applications

In Sect. 1.7 a total world sales of at least 0.5 million US dollars annually were set as the criterion for classifying a biocatalyst as "industrial." As shown in the present chapter, the current annual world sales of some of the biocatalysts under specific consideration are lower (e.g., immobilized ß-galactosidase). As we know, immobilization makes it possible for biocatalysts to be used repeatedly or continuously so that large, even immense quantities of product can be obtained with small or only minute amounts of biocatalyst. It thus seems justifiable to classify a process as being "industrial" or "nonindustrial" on the basis of the scale of the production process carried out with the help of immobilized biocatalysts rather than based on the value of the catalysts themselves. In the following, a process is termed "industrial" if practiced in a factory on the scale of a daily output of at least several tons of product (day tons = dato).

Despite the fact that biocatalysts can be used repeatedly or continuously in their immobilized form, they still, in many cases, cannot compete successfully (from the economic-technical point of view) with native biocatalysts and other chemical procedures. A variety of criteria play a role in such considerations and some of the important factors favoring the use of immobilized biocatalysts are the following:

> A great demand for the product
> Favourable reaction kinetics
> A high market price for the product
> A low-molecular substrate
> A nonturbid substrate solution
> Native biocatalysts not already in use
> Freely dissociating coenzyme not required
> The need for a pure product
> High price of the native form of the biocatalyst
> No competitive chemical procedure

The importance attaching to the individual factors will differ for each biocatalyst and in each concrete case since requirements vary enormously depending on the product and on the type of production plant. For this reason no attempt has been made to arrange the above factors according to importance.

5.1 Classical Fields of Application

Long before immobilization had become a topic of planned research, immobilized microorganisms were being employed industrially on a purely empirical basis in the microbial production of vinegar using wood shavings overgrown with bacteria, and in the trickling filter or percolating process for waste water clarification. Although in the meantime submerged techniques have gained increasing ground in both of these classical fields, the basic principles of these trickling filter method warrant a brief description.

Trickling Processes for Vinegar Production

In the production of vinegar by the trickling process, an alcohol-containing mash is allowed to trickle over beech or birch shavings on which bacteria are growing (see Fig. 58). The acetic acid bacteria adhering (immobilized) to the wood shavings oxidize the ethanol to acetic acid with the help of the oxygen contained in the air and directed through the reactor contents against the flow of mash.

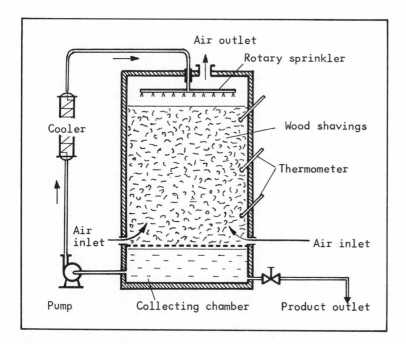

Fig. 58. Scheme of a generator for acetic acid production

The older trickling processes (i.e., the Schuezenbach process) were carried out in relatively small reactors with a volume of about 2 m^3. Later, in the so-called generator process, large wooden vessels ("generators") with volumes up to 60 m^3 were used. Generators of this type, also known as Frings generators after the manufacturing firm of Frings (Bonn, West Germany), enjoyed their greatest popularity from 1925 to 1955; some of them are still in use today. For new installations, however, submerged procedures are given preference on account of their higher volumetric productivity.

Vinegar production in the generator shown in Fig. 58 is a semicontinuous process. The mash filling the collecting chamber (wine with approximately 12 % alcohol and added nutrients) is pumped continuously through the apparatus for 3 days at a temperature of 26° C to 30° C. The rotating sprinkler on the top of the generator ensures an even flow of the mash over the shavings and the attached microorganisms. To guarantee an adequate oxygen supply, air is blown into the carrier column from the bottom. As soon as the alcohol content has sunk to about 0.3 vol% due to acetic acid formation the product is drawn off and the mash renewed.

The yield of acetic acid from the ethanol substrate is usually around 85 % to 90 % of the theoretical value. Small losses occur due to the requirements of the bacteria themselves, and due to evaporation of the ethanol. On the whole, the operation of the generators is uncomplicated. The low pH value, the adequate supplies of oxygen from the air, of acetic acid and alcohol favor the acetic acid bacteria to such a degree that contaminations due to other microorganisms are almost unknown. The generators can therefore be operated for years with one and the same set of shavings.

Percolating Procedures in Waste-water Treatment

Percolating or trickling filter processes for the aerobic treatment of waste water are very similar to the trickling process for the production of vinegar. The clarification procedure, first brought into use at the end of the last century, is based on a solid bed reactor known as a trickling filter. The filter consists of slag or porous synthetic materials on which colonies of microorganisms have been allowed to settle (immobilize) in order to bring about aerobic breakdown of the organic matter present in the waste water.

The organisms colonizing the porous material constituting the filter are chosen according to the nature of the water to be clarified; the communities are variously composed of bacteria, fungi, protozoa and even higher organisms, such as small crustaceans, worms and insect larvae. The microorganisms are largely responsible for breaking down the impurities in the water, whereas the higher organisms prevent clogging of the filter by breaking down the biomass accumulating on the porous materials.

The installation shown schematically in Fig. 59 is usually 3 to 4 m in height and 10 to 30 m in diameter. It is filled with lumps of coarse-pored material, such as slag, volcanic material or plastic, supported on a grid about 0.5 m from the bottom. Side openings in the wall ensure the circulation of air, thus providing the setup with oxygen. In winter, when the temperature of the waste water exceeds that of the outside air, the air rises through the filter from below. On very hot summer days the situation can be reversed: the air is cooled by the waste water, becomes heavier than the air outside, and flows downward through the reactor.

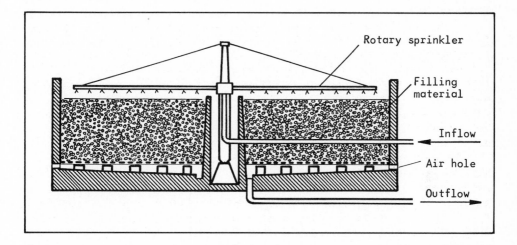

Fig. 59. Trickling filter for waste-water treatment

Many modifications of the classical trickle filter are in use. One alternative to the form shown in which the waste water is sprinkled onto a solid bed (see Fig. 59) is to use plates or hollow cylinders which can be alternately submersed in the waste water or raised in the air. The two methods, the trickling filter and the submersed body procedure, have in common that the organisms exploited for water purification are confined in or on porous solid material. In other words both methods employ complex, immobilized biocatalysts.

5.2 Production of L-Amino Acids Using L-Aminoacylase

The first industrial application of an immobilized enzyme resulting from the intensive research on methods of immobilization was introduced in 1969 for the production of L-amino acids using carrier-bound L-aminoacylase (= L-aminoamidase). The process, developed and applied on a commercial scale by the Japanese firm of Tanabe Seiyaku, exploits the ability of L-aminoacylase to break down specifically only the L form of acetylated amino acids, after which, as free acids, they can easily be separated from the acetylated D form by crystallization.

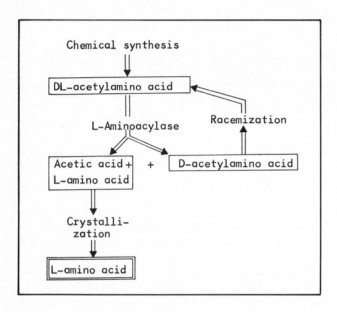

Fig. 60. Scheme of the L-aminoacylase process

The scheme in Fig. 60 shows the basic steps in the production of L-amino acids using L-aminoacylase. A racemic mixture of D- and L-amino acids can be produced relatively easily by chemical synthesis, and is then acetylated by chemical means. This is followed by deacetylation of only the L-acetyl amino acid by the ionically bound L-aminoacylase. The physiologically more important L-amino acid can then be crystallized out of the mixture and is thus obtained in the pure form. The D-acetyl amino acid is now racemized (transformed into a mixture of D and L forms by a simple chemical process), and once more the L-acetyl amino acid is split by the enzyme and the L-amino acid is crystallized out. Thus, in the end, all of the chemically synthesized DL-amino acid is transformed into pure L-amino acid.

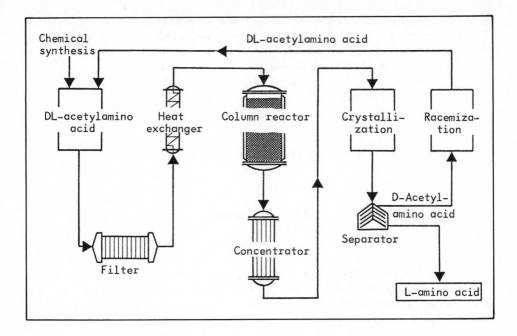

Fig. 61. Scheme of the production of L-amino acids using ionically bound
L-aminoacylase

That the above procedure, illustrated in Fig. 61, is profitable despite
the high cost of the carrier material Sephadex is chiefly due to the fact
that the DEAE Sephadex can be reused for years on end. Its regeneration,
i.e., the recharging of the carrier material with active L-aminoacylase,
is simple and is performed at regular intervals of roughly five working
weeks. During this time the activity of the carrier-bound enzyme sinks to
about 60 % to 70 % of its original value. The half-life of the L-aminoacylase (from *Aspergillus oryzae*) bound to the ion exchanger is about 10
to 12 working weeks.

As Fig. 61 shows, the bound L-aminoacylase is employed in a columnar
bed reactor with a volume of about 1 m^3; the operating temperature is 50°
C. The specific flow rate of the substrate through the reactor, also
termed the space velocity, is of the order of 0.5 to 2 h^{-1}, depending on
the L-acetylamino acid to be broken down. The specific flow rate is given
by the ratio of the flow rate (e.g., in 1/h) to reactor volume (e.g., in
1). When using data from the literature it should be taken into consideration that in many cases only part of the total volume of the reactor
is included in the author's calculations (e.g., the free fluid volume or
the volume of the biocatalysts), which means that the flow rates and volumetric productivities calculated in this way are higher than if the total
volume had been taken.

The racemate separation method using L-aminoacylase is mainly employed
for producing the amino acids L-methionine, L-phenylalanine and L-valine
on an annual scale of several hundreds of tons of each. The production

costs are roughly half those of the earlier discontinuous process in which soluble enzyme was used. The chief saving is due to the fact that the L-aminoacylase, instead of being lost after each batch, can be used continuously for long periods of time.

5.3 L-Amino Acid Production in Membrane Reactors

The chemical synthesis of amino acids competes not only with microbial and enzymatic processes but also with extraction of the acids from natural proteins. Very recently, especially good progress has been made in the enzymatic procedures. West German research teams headed by C. Wandrey and M.R. Kula, in cooperation with the Degussa concern, have succeeded in developing on an industrial scale membrane reactors incorporating several enzymes and coenzymes. This opens up attractive alternatives to the methods so far available.

Amino Acids from Keto Acids

L-amino acids can be produced from α-keto acids by stereoselective reductive amination with L-amino acid dehydrogenase. Figure 62 shows the general scheme of this reaction which involves consumption of the coenzyme $NADH_2$ (more accurately $NADH + H^+$). Regeneration of the coenzyme can be satisfactorily achieved by coupling the amination with enzymatic oxidation of formic acid to carbon dioxide. NAD (i.e., NAD^+) is reduced back to $NADH_2$ in the course of the reaction catalyzed by formate dehydrogenase.

In a continuous process, already in operation on an industrial scale, the two apoenzymes and the coenzyme are put to use in membrane reactors. As well as the two high-molecular enzymes, the low-molecular coenzyme has to be retained by the membrane; for this purpose a volume increase is achieved by coupling the coenzyme with water-soluble polymers, particularly polyethylene glycol (PEG), with molecular weights in the range of 10,000 to 20,000.

The pharmaceutically important amino acids L-methionine, L-valine and L-phenylalanine are already being produced on a scale of about 200 tons annually, using a procedure based on the above principle. The process as carried out in membrane reactors appears to be cheaper than racemate separation with L-aminoacylase.

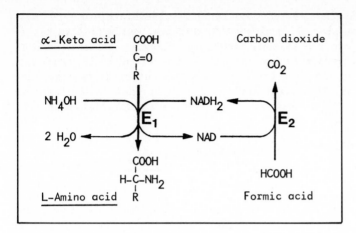

Fig. 62. Reactions to produce L-amino acids from α-keto acids.
E_1 amino acid dehydrogenase; E_2 formate dehydrogenase

Amino Acids from Hydroxy Acids

A variant of the production of amino acids from α-keto acids involves a preliminary step in which α-hydroxy acids are converted into α-keto acid. This eliminates the necessity for cofactor regeneration by the formate dehydrogenase since the regeneration takes place during the oxidative conversion of hydroxy acids into keto acids. The production of alanine as shown in Fig. 63 illustrates the principle of the method.

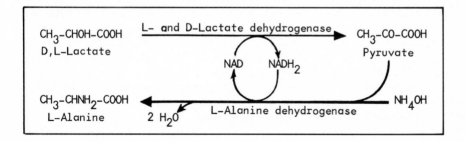

Fig. 63. Scheme of the L-alanine production

The starting material for the production of L-alanine (see Figs. 63 and 64) is a synthetic (and cheap) racemic mixture of L- and D-lactate. By using L- and D-lactate dehydrogenase both enantiomeric forms of the lactate are converted to pyruvate with the reduction of NAD. In the subsequent conversion of pyruvate to L-alanine, catalyzed by L-alanine dehydrogenase, the NADH$_2$ (in its enlarged form) is regenerated to NAD.

In addition to the starting substance DL-lactate, pyruvate also has to be fed continuously into the system, albeit in minute quantities. This is due to the fact that, before it can be further converted to L-alanine, a very small fraction of the pyruvate arising as an intermediate from lactate passes through the ultrafilter membrane. Without small additions from outside the pyruvate concentration would in time approach zero and the reaction would come to a stop.

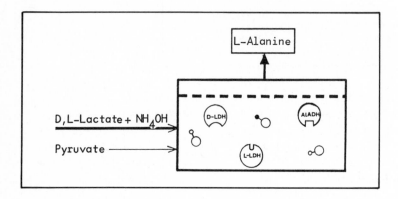

Fig. 64. Multi-enzyme membrane reactor for the production of L-alanine. **ALADH** alanine dehydrogenase; **D–LDH** D-lactate dehydrogenase; **L–LDH** L-lactate dehydrogenase

At present, the relatively cheap L-alanine is not produced on a larger scale by the membrane reactor method. The significance of the method lies in the fact that it offers a model for the efficient use of membrane-separated multiple enzyme systems with coenzyme regeneration.

5.4 The Production of Fructose-containing Syrups

The sweetening properties of fructose are considerably greater than those of glucose. This is the principle reason for the isomerization to fructose of part of the immense quantities of glucose produced from corn and other starch varieties. As shown in Fig. 65, the isomerization is catalyzed by the enzyme glucose isomerase until a state of equilibrium is attained at about 50 % each of fructose and glucose.

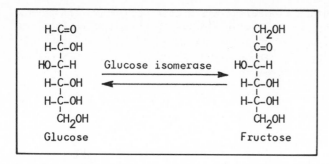

Fig. 65. Isomerization of glucose to fructose

Since the late 1960's glucose isomerase has been in use for the industrial production of fructose-containing syrups. These syrups are marketed under diverse trade names, as isosyrups or under the abbreviation HFCS (high fructose corn syrup). Although soluble glucose isomerase was used earlier, today the immobilized form of the enzyme is almost exclusively employed. The organisms used in the production of glucose isomerase and the immobilization techniques vary widely from one firm to another. Some examples are given in Table 18.

Table 18. Commercial immobilized glucose isomerases

Trade name	Producer	Enzyme source	Method of immobilization
Maxazyme GI	Gist Brocades, Netherlands	*Actinoplanes missouriensis*	Co-crosslinking of enzyme and gelatine
Optisweet	Miles-Kali-Chemie, Fed. Rep. Germany	*Streptomyces rubiginosus*	Covalent coupling to silanized ceramics
Sweetase	Nagase Biochemical, Japan	*Streptomyces phaechromogenes*	Cross-linking of whole (dead) cells
Sweetzyme	Novo Industri, Danmark	*Bacillus coagulans*	Cross-linking of whole (dead) cells
Spezyme	Finnsugar Ltd., Finland	*Streptomyces rubiginosus*	Ionic binding to polystyrene coated DEAE-cellulose
Takasweet	Miles, USA	*Microbacterium arborescens*	Cross-linking of whole (dead) cells

All glucose isomerases so far known are relatively insensitive to temperature; their pH optimum is in the neutral to slightly alkaline range. They require bivalent metallic ions (Co^{2+}, Mn^{2+} or Mg^{2+}) as cofactors but are inhibited by other bivalent ions, especially Cu^{2+}, Zn^{2+} and Hg^{2+}, as well as by the sugar alcohols sorbitol and xylitol. They are rapidly inactivated in the presence of oxygen.

Strictly speaking, glucose isomerases are xylose isomerases and have a very much higher affinity (lower K_m value) for aldopentose xylose than for aldohexose glucose. The structure of xylose is almost identical with that of glucose (it contains only one HCOH group less), which is why the enzyme mistakes one substrate for the other.

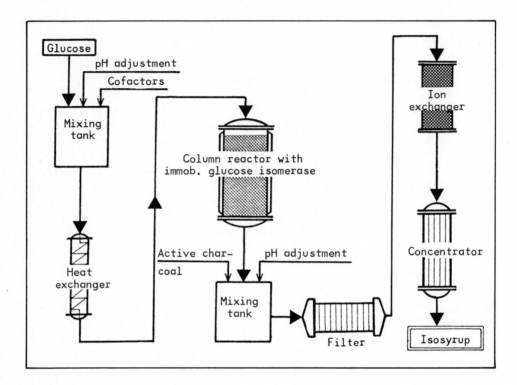

Fig. 66. Production process for fructose rich syrups

Technically (see Fig. 66), immobilized glucose isomerase is usually employed in continuously operated packed- and fluidized-bed reactors at pH values between 7.5 and 8.0, and at temperatures of about 60° C. Glucose is used in a 40 % to 50 % solution. The fructose-containing product is finally concentrated to a dry weight content of 70 % to 75 %. The syrups on the market normally contain about 42 % fructose (dry weight), in addition to roughly 52 % glucose and 6 % oligosaccharides. Such syrups have about the same sweetening properties as normal household sugar (sucrose).

Under industrial production conditions, immobilized glucose isomerases have half-life values in the order of 1000 working hours. The gradual drop in activity is compensated by reducing the flow rate or by linking together several reactors whose enzyme contents are of different ages. As a rule, the enzyme contents of a reactor are renewed after 2000 to 3000 working hours (see Fig. 67). A total of about 2 to 3 tons of isosyrup can be produced with 1 kg of the commercially available immobilized glucose isomerases (see Table 18), although some (more expensive) preparations with a specially high activity give as much as 20 tons of product per kg enzyme.

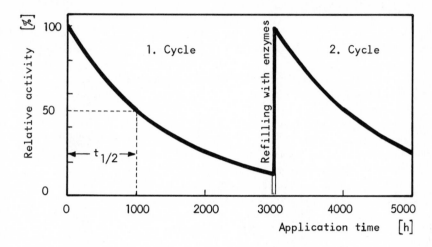

Fig. 67. Activity of an immobilized glucose isomerase in the course of an industrial application

In Europe, fructose-containing syrups play a much smaller role than in Japan and the USA, largely due to quotas of the European Economic Community. The annual world market at present is about 6 to 7 million tons of isosyrup, for the production of which 1000 to 2000 tons of immobilized glucose isomerase are required. Thus, glucose isomerase is by far the most important enzyme employed in the immobilized form.

5.5 Derivatives of Penicillins

6-Aminopenicillanic acid (6-APA) is the starting substance for the production of almost 20 different semisynthetic penicillins. Since the yields of 6-APA obtained from native penicillins by microbial fermentation or by the chemical removal of side chains are unsatisfactorily low, the most popular method nowadays employs immobilized penicillin acylase (= penicillin amidase) to split off the phenyl acetic acid side chain from benzyl penicillin (penicillin G) obtained by fermentation. Figure 68 shows the hydrolytic removal of the phenyl acetic acid side chains as catalyzed by penicillin acylase.

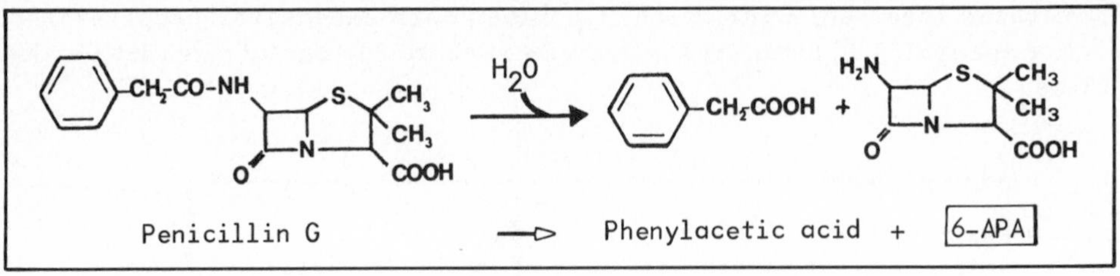

Fig. 68. Removal of the side chain from penicillin G by means of penicillin acylase

The reaction shown in Fig. 68 is reversible: the breakdown reaction proceeds at neutral to slightly alkaline pH values (7 to 8), whereas under acid conditions (pH 4 to 6) the reaction is predominantly synthetic. The enzymatic addition of new side chains to 6-APA by means of penicillin acylase also finds occasional use. One example, shown in Fig. 69, is the production of ampicillin, and involves the coupling of phenylglycine methylester to the penicillin nucleus. The penicillin acylase used for this purpose can be isolated from *Pseudomonas* species, it should more correctly be termed ampicillin acylase.

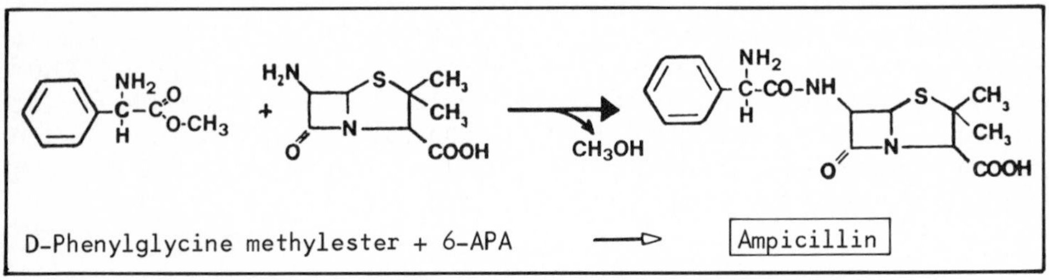

Fig. 69. Production of ampicillin by enzymatic coupling of the side chain to 6-APA

Penicillin acylase occurs in a whole series of fungi as well as in many bacteria. For the industrial production of 6-APA penicillin acylase from *Escherichia coli* is usually employed either bound to Sephadex or some other carrier, or in the intracellularly bound form. Occasionally, even native cells of *E. coli* are used. It is important for the preparations to be entirely free of penicillinase (= ß-lactamase) since its action would split the ß-lactam ring, thus destroying the active penicillin nucleus. Reactions with immobilized penicillin acylase are usually carried out batchwise in a stirred reactor at 35° C to 40° C as shown in Fig. 70. At the end of the reaction the immobilized enzyme is separated for reuse in the next batch. In this way, 150 to 200 kg of 6-APA can be produced with only 1 kg of enzyme. The pH has to be held at a constant level of 7 to 8 by means of controlled additions of alkali. Without this adjustment the phenyl acetic acid set free would rapidly lower the pH to a value at which the enzyme would primarily catalyze the synthetic reaction, or even be inactivated. For this reason the packed bed is unsuitable for such re-actions. However, one way of circumventing this difficulty is to combine packed bed and stirred tank reactor, the adjustments being made in the stirred tank. In this variant the substrate is circulated rapidly under pressure through the enzyme-filled packed bed and continuously returned to the stirred reactor. The drop in pH within the packed bed can thus be kept within very close limits and the separation of the enzyme particles from the product is simplified.

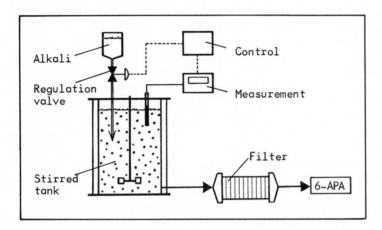

Fig. 70. Example of a process for producing 6–APA

5.6 Production of Aspartic Acid

L-aspartic acid is used in medicine and as a food additive. It can be prepared either by microbial fermentation or from fumaric acid and ammonia using immobilized L-aspartase, according to the following reaction:

$$HOOC–CH=CH–COOH \quad + \quad NH_3 \quad \xrightarrow{\text{L–Aspartase}} \quad HOOC–CH_2–\underset{\underset{NH_2}{|}}{CH}–COOH$$

Fumaric acid   +   Ammonia   L-aspartic acid

The method developed by I. Chibata (Tanabe Seiyaku, Japan) has been in practical use since 1973. It employs the intracellular L-aspartase of *E. coli* cells trapped in a matrix of polyacylamide or carrageenan. This method is often cited as the first known procedure in which whole immobilized cells were used. It should be added, however, that neither living cells nor multistep reactions are involved. In fact, the one enzyme that is used is left within the cell structure for the purely economic and practical reasons described below.

Extraction of L-aspartase and its subsequent coupling to carriers has proved to be unsatisfactory since the isolated enzyme is highly unstable once outside the cell, and a considerable amount of the activity is lost during the preparation. An appreciable increase in the activity of the cell-bound aspartase is achieved by partial autolysis of the cells. Apparently the breakdown of the membrane barrier results in better mass transfer and consequently in a higher turnover.

In the industrial preparation of aspartic acid a 1 M solution of ammonium fumarate is used as substrate in column reactors of approximately 1 m^3 volume. The stability of the biocatalysts is improved by the addition of 0.1 mmol Mg^{2+} per liter of substrate. The operation is carried out at a specific flow rate of roughly 0.6 h^{-1}, a temperature of 37° C and a pH of 8.5. The result is a daily output per reactor of about 2 tons of aspartic acid. Extraction of the product from the outflow is simple: the pH is lowered to 2.8, the isoelectric point of L-aspartic acid, and the temperature to about 7. The product precipitates as crystals and can be separated by filtration.

In the above procedure using dead immobilized *E. coli* cells, about 95 % of the theoretical yield is obtained. The half-life of the aspartase is approximately 120 days, whereas native cells have a half life of roughly 10 days. The higher stability of the immobilized aspartase means a reduction in costs, particularly those connected with the microbial production of the enzyme. By using the immobilized system instead of native cells an overall saving of about 40 % is achieved.

5.7 Application of ß-Galactosidase

As shown in Fig. 71, ß-galactosidase (= lactase) breaks down the milk sugar lactose into the monosaccharides glucose and galactose, which taste sweeter and crystallize out less readily than lactose. Many people, especially among the negro population, are unable to tolerate lactose because they suffer from a deficiency of the enzyme lactase. As a consequence, there is a large field of application for ß-galactosidase in the breakdown of lactose in milk and milk products and the preparation of sweet syrups from whey, which contains large amounts of lactose.

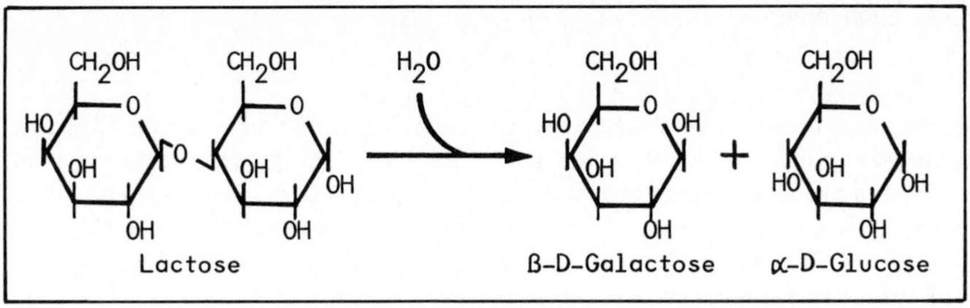

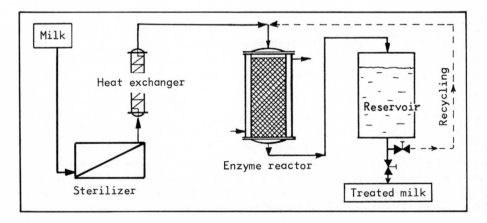

Fig. 71. Enzymatic breakdown of lactose to glucose and galactose

Since ß-galactosidases are relatively expensive as compared with other technical enzymes, numerous attempts have been made to utilize them for industrial purposes in an immobilized form. So far only partial successes have been registered in what is potentially a vast field of application.

Lactose Hydrolysis in Milk

Milk is a difficult substance for treatment with immobilized enzymes due to the fact that it consists of an emulsion of fine droplets of fat in a watery medium. Emulsions of this nature cannot normally be treated in a packed bed without the danger of separation and clogging. However, by using yeast lactase spun into cellulose acetate fibers the Italian firm of Snam Progetti has developed a practicable procedure for the treatment of milk in packed bed reactors (see Fig. 72). The enzyme-containing fibers, even when densely packed, lead neither to undesirable pressure buildup nor to separation of the milk emulsion or clogging of the packed bed. The procedure is carried out on a scale of about 10 tons daily in the dairy plant in Milan.

Fig. 72. Process for hydrolysing lactose in milk

Sweet Syrups from Whey

For many years attempts have been made to convert lactose from surplus whey into sweet syrups with the help of immobilized ß-galactosidase. In general, ß-galactosidases of mold origin (*A. niger, A. oryzae*) are used because their pH optimum is better suited to the acid whey than that of the yeast enzyme. To date, the use of such procedures in industry has been limited to isolated experiments. The main reasons why immobilized ß-galactosidases have not achieved greater popularity in this field are, firstly, that they are inhibited to an appreciable degree by their reaction product galactose, and secondly, as can be seen from Table 19, they have only a poor half-life in original and deproteinized whey.

Table 19. Typical values for the half-life of ß-galactosidase in different media

Substrate	Half-life
Acid whey (original)	6 days
Deproteinized whey	10 days
Deionized + deproteinized whey	60 days
5 % Lactose solution	90 days

5.8 Further Possible Industrial Uses

The most important current fields of application for immobilized biocatalysts have been considered in the preceding Sects. 5.1 to 5.7. A few other fields that have already achieved or may soon achieve industrial status are briefly described in the following.

L-Malic Acid from Fumaric Acid

Fumaric Acid, which is easily prepared by chemical synthesis, besides its use as starting material in the production of L-aspartic acid as described in Sect. 5.6, can also be used to prepare L-malic acid. The reaction is catalyzed by the enzyme fumarase and proceeds according to the equation

$$HOOC-CH=CH-COOH \quad + \quad H_2O \quad \xrightarrow{\text{Fumarase}} \quad HOOC-CHOH-COOH$$

The reaction as first introduced on an industrial scale by Tanabe Seiyaku (Japan) in 1974 employed dead cells of *Brevibacterium ammoniagenes* embedded in polyacrylamide. In the meantime, cells of *Brevibacterium flavum* embedded in k-carrageenan are preferred on account of their longer half-life. Prior to being embedded, the cells are exposed to solvents to enhance their permeability; this facilitates the washing-out of coenzymes and thus prevents the undesirable formation of succinic acid via cofactor-dependent side reactions.

The industrial reaction is carried out at pH 7.5 and a temperature of 27° C, in 1 m^3 reactor columns with specific flow rates of approximately $D = 0.3$ h^{-1}. The substrate employed is 1 M sodium fumarate, which is converted to about 70 % into malic acid. The half-life of the system is roughly 160 days.

Very recently, pilot experiments have also been carried out in enzyme membrane reactors using purified fumarase from *Brevibacterium ammoniagenes*. A disadvantage of the fumarase reaction is the inhibition of the enzyme by the product malic acid. This causes more serious disturbance in the homogeneous mixture of membrane reactors than in packed bed reactors with plug flow behavior. In an attempt to decrease the product inhibition, a two-step procedure with partial removal of the enzyme from the reactor is being considered.

On the whole, the prospects for malic acid production with immobilized biocatalysts are limited, due to strong competition from microbial fermentation and chemical synthesis.

Raffinose Hydrolysis in Sugar Factories

Raffinose, which is present in sugar beet in small quantities (about 0.1 %), becomes concentrated during the process of sugar manufacture and has an inhibiting effect on the crystallization of the sucrose. The raffinose can be broken down into sucrose and galactose by α-galactosidase (= melibiase), as shown in Fig. 73, and the yield of beet sugar thus increased. The α-galactosidase must be entirely free of invertase activity since this would break down the desired end product of sugar manufacture, sucrose, into fructose and glucose.

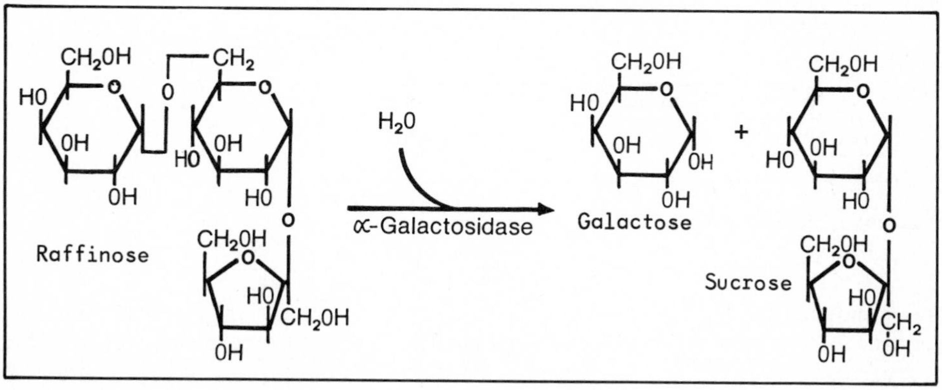

Fig. 73. Break down of raffinose by α-galactosidase

Certain organisms used for other industrial purposes are able to produce α-galactosidase. It occurs in bottom-fermenting beer yeast, which is nowadays allocated with the non-α-galactosidase-producing distiller's-, wine- and baker's-yeast to the species *Saccharomyces cerevisiae*. Formerly, the bottom-fermenting beer yeasts were considered to be a separate species on account of their ability to synthesize α-galactosidase. Another micro-organism capable of forming this enzyme and widely used in industry is *Aspergillus niger*. Both organisms have been considered as sources of α-galactosidase, but the high degree of invertase activity associated with both organisms posed too many difficulties.

More promising developments have been reported by the American Great Western Sugar Company with *Mortierella vinacea*, and by the Japanese Nippon Beet Sugar Company with *Absidia* species. In both cases dead, cross-linked mycelia are used in a continuous procedure at temperatures of about 50° C.

The Use of Amylolytic Enzymes

Amylolytic enzymes in soluble form occupy a firmly established position in the starch industry and their sales are correspondingly high (see Sect. 1.7). Figure 74 shows the three most important enzymes used in the starch industry. We have already encountered glucose isomerase in Sect. 5.4, as the most important of all immobilized enzymes. Of the other two enzymes, α-amylase can only be employed in soluble form since the molecular weight of its substrates amylose and amylopectin are too high for satisfactory hydrolysis with immobilized enzymes.

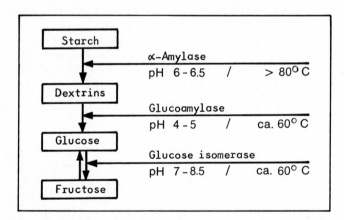

Fig. 74. Enzymes used in starch processing

The importance of the conversion of starch to sugar has stimulated in-numerable attempts to find a corresponding immobilized preparation as a substitute for soluble glucoamylase (= amyloglucosidase). Nevertheless, no striking industrial success has so far been reported, the reasons for this being the unsatisfactory heat stability and activity characteristics of the immobilized enzyme. For example, with immobilized glucoamylase it is impossible to obtain the DE values achieved with the soluble form of the enzyme. Further reasons for the lack of interest in the immobilized form are that the existing procedure is so well established, and the soluble preparation is very cheap.

The prospects for immobilized glucoamylase may well be more favorable in other fields than the starch industry since totally different criteria are involved. To suggest only one example: the use of immobilized gluco-amylase for the production of diet beer in preference to soluble enzyme would guarantee the absence of enzyme protein.

Biogas Production

In waste-water technology microorganisms have been put to practical use in the aerobic trickling filter procedures (cf. Sect. 5.1) and in some of the more modern anaerobic forms of waste treatment in use since 1970. Two developments of this kind, shown in Fig. 75, have made possible an efficient production of biogas by using biomass confined in a reactor.

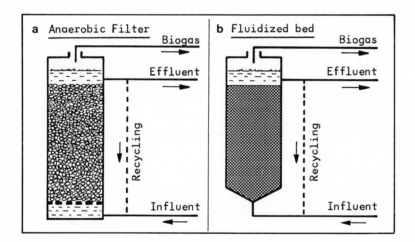

Fig. 75a, b. Anaerobic filter and fluidized bed for biogas production

In both types of reactor shown in Fig. 75, the waste water flows through the reactor from below, the biogas being produced by bacteria bound to solid material. Highly loaded waste water can be diluted by reentering some of the already purified effluent. On account of the danger of clogging, such systems are unsuitable for the treatment of waste water containing solid particles. Like the familiar aerobic trickling filter, the anaerobic filter contains coarsely pored carrier material, such as coke, slag or porous plastic. The methanogenic bacteria settle in and on the carrier material (dimensions: a few cm).

Anaerobic filters are well suited to the treatment of waste water containing large amounts of carbohydrate. Normally the procedure can be used up to a chemical oxygen demand (COD) of 10 g/l, but higher COD values soon lead to clogging on account of the excessive increase in biomass. In such cases, the COD values of the inflow can be lowered by partial reentry of the clarified effluent. At a volumetric COD load of 2 to 4 kg/m^3 as much as 90 % removal of the COD can be achieved. In the presence of methanol, formic acid and acetic acid, which are direct substrates for methane formation, 90 % breakdown can be achieved at volumetric COD loads of up to 20 $kg/m^3 \cdot d$.

The fluidized bed illustrated in Fig 75b functions in a way similar to the anaerobic filter, with the difference that the particle size of the solid bodies in the fluidized bed reactor never exceeds 1 mm. This results in a very large specific surface area, i.e., roughly 300 m^2/m^3 and hence in a very high performance, which can amount to as much as 90 % breakdown at volumetric loads of 10 kg/m^3, or even more.

In addition to the two systems shown in Fig. 75 there are several other reactor forms in which retention of biogas-producing bacteria is practiced. Of this type, the most widely used on an industrial scale is the so-called UASB reactor (upward anaerobic sludge blanket). In this procedure the waste flows upward through a thick layer of flocculating bacteria. No carrier material is required, but the bacteria can nevertheless be regarded as a special form of immobilized biocatalyst because the biomass is held back (immobilized) by artificial means such as built-in baffles or by the addition of flocculating agents.

6 Application in Analytical Procedures

6.1 Affinity Chromatography

As well as being a valuable analytical method, affinity chromatography can also be used to extract pure components from complex mixtures of substances. The method exploits the property of certain reactants to recognize each other and to form complexes. Affinity of this kind exists between antigens and antibodies, hormones and receptor proteins, to mention only two examples. As Table 20 shows, enzymes, too, have a specific affinity for a variety of substances.

Table 20. Substances with specific affinity to enzymes

– Substrates of the enzyme	– Enzyme inhibitors
– Products of the enzyme	– Antibodies
– Coenzymes	– Allosteric effectors

In affinity chromatography, one of the two partners capable of interaction is rendered insoluble either by coupling it (the reactive ligand) to a carrier (adsorbent) or by cross-linking. If a complex biological mixture containing a substance with a specific affinity for the ligand is allowed to flow over the adsorbent in a column, only this particular substance will be adsorbed. A column filled with particles of an immobilized enzyme, for example, will only retain a substance which has an affinity for this enzyme. The greater its affinity for the ligand in the chromatographic column, the longer is the "target" substance retained in it (see Fig. 76). It can be removed, i.e., eluted, from the column by adjusting the physical conditions such as pH and temperature, or by controlled addition of other substances with an even higher affinity for the ligand.

As Table 20 shows, a wide variety of substances possess a specific binding capacity for enzymes; in their immobilized form, therefore, enzymes are in widespread use in analysis and purification procedures. Conversely, an immobilized substrate can be used to purify or separate an enzyme from mixtures, although in this case the substrate must be denatured in such a way that it is still recognized by the enzyme but no longer broken down by it. This is achieved in the case of certain polysaccharides, for example, by cross-linking with epichlorhydrin. A simpler method than using cross-linked substrate is to employ immobilized inhibitors which cannot react with the enzyme. The principle involved in separation by affinity chromatography and the typical phases in the procedure are shown in Fig. 76.

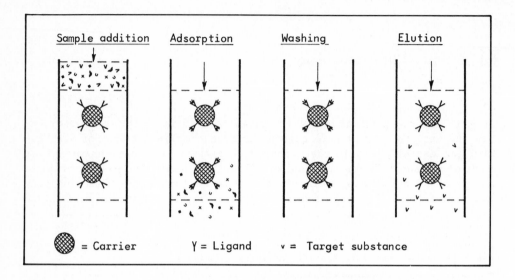

Fig. 76. Principle and procedure of an affinity chromatography

Substances used as carriers in affinity chromatography have to conform to certain standards: they must combine high chemical and mechanical stability with the necessary physical properties, such as porosity and stability under pressure, they should couple readily with the ligand (e.g., enzymes), and not enter into undesirable adsorption reactions. Carriers in common use are agarose (Sepharose[R]), dextran (Sephadex[R]) and polyacrylamide (Bio-Gel[R]), all of which are also commercially available in activated forms, i.e., forms that readily couple with specific groups (e.g., $-NH_2$) on the ligand.

6.2 Automatic Analyzers

By allowing a substance to react with the appropriate immobilized biocatalyst and registering the accompanying changes, it is possible to arrive at a quantitative determination of the original substance. The high degree of specificity of biocatalysts makes it possible to analyze the most complex mixtures. In a typical example (see Fig. 77), a sample is entered into the analyzer by means of a dispenser and is then pumped through a minireactor containing immobilized enzyme. Finally, if required, after additional reactions (e.g., with color reagents), the sample is analyzed as it flows through the spectrophotometer, fluorometer, polarimeter or whatever type of analyzer is installed.

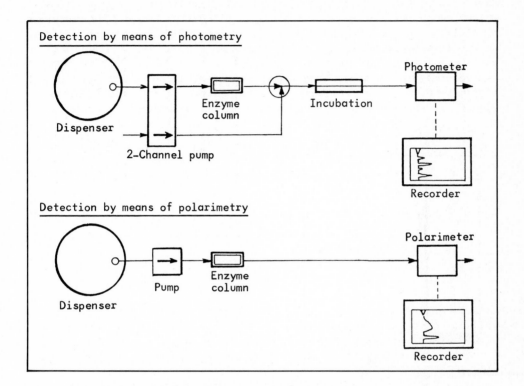

Fig. 77. Scheme of two autoanalyzer types using different methods to detect the products of the enzymatic reaction

In automatic analyzers of the type shown schematically in Fig. 77 the immobilized enzymes are usually packed in particulate form in a mini-column, or are covalently bound to the inner wall of coils of nylon tubing or of glass tubes. To eliminate the danger of backmixing, each treated sample, as it flows through the tubing coils, is sealed by air bubbles.

6.3 Biochemical Electrodes

In biochemical electrodes, also known as biosensors or bioelectrodes, the sample component to be analyzed is converted specifically by one or more enzyme reactions. The biocatalysts act as biochemically specific receptors that alter the substrate in a particular way. This change is recorded by an electrical transducer which, in the case of biochemical reactions, is usually a gas- or ion-sensitive electrode.

Depending on whether single enzymes or whole cells are used in the biocatalytic reaction, we speak of either enzyme or microbial electrodes. Figure 78 shows the typical construction of a biochemical electrode.

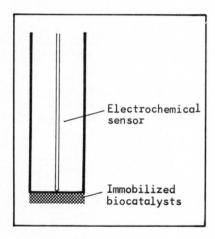

Electrochemical sensor

Immobilized biocatalysts

Fig. 78. Scheme of a biochemical electrode

For a biochemical electrode to function efficiently, it is important for the immobilization of the biocatalyst to be carried out in the immediate surroundings of the sensitive region of the electrode. Immobilization itself can be performed in a variety of ways, as for example, by direct coupling to the electrode surface, by embedding in a gel, or by encapsulation within a membrane.

Simple glass electrodes registering changes in H^+ concentration, i.e., indicating pH values, are often used. Of the more than 20 types of electrodes sensitive to different ions, those reacting to ammonium-, phosphate- or iodide ions are the most commonly used in biochemical electrode systems. The gas-sensitive oxygen and carbon dioxide electrodes are also very popular. Table 21 lists some examples of substances which can be measured with biochemical electrodes.

The examples in Table 21 are only a small selection from the many possibilities so far known. For each of the substances listed in the Table, only one of its possible methods of detection is given, whereas in fact several alternatives are often available, both as regards the biocatalyst and the electrochemical sensor employed.

Table 21. Examples of biochemical electrode systems

Substance	Immobilized biocatalyst	Type of electrode
L-Amino acid	L-Amino acid oxidase + peroxidase	I^--sensitive elektrode
Cholesterin	Cholesterin oxidase	O_2-electrode
Glutamic acid	*Escherichia coli* (cells)	CO_2-electrode
Urea	Urease	NH_4-sensitive electrode
Lysine	Lysine decarboxylase	CO_2-electrode
Penicillin	Penicillinase (ß-Lactamase)	Glass electrode
Phenol	*Trichosporon cutaneum* (cells)	O_2-electrode
Hydrogen peroxide	Catalase	O_2-electrode

Missing from Table 21, however, is any mention of the extremely important biochemical electrodes for the measurement of glucose. The use of enzyme electrodes has often been suggested for the specific, quantitative determination of glucose in fermentation processes or in body fluids, etc. Most electrodes used in this connection contain the enzyme glucose oxidase, which reacts specifically with ß-D-glucose but also measures the ∝-D-glucose since this is converted into the ß form by mutarotation. The enzyme reaction

$$\text{ß-D-Glucose} + O_2 + H_2O \longrightarrow \text{Gluconic acid} + H_2O_2$$

can be electrochemically detected and quantified in a number of ways. For example, the drop in pH resulting from the formation of gluconic acid can be detected with a glass electrode, or, which is almost as simple, the oxygen consumed in the reaction can be measured with an O_2 electrode. A common practice is to allow the H_2O_2 to undergo a further reaction which is then registered:

$$H_2O_2 + 2\ I^- + 2\ H^+ \longrightarrow 2\ H_2O + I_2$$

This reaction of the hydrogen peroxide with iodide, catalyzed by the enzyme peroxidase results in a drop in the iodide content of the micro-environment of the electrode tip and can therefore be measured with an iodide-sensitive electrode.

A serious problem that prevents the more universal use of glucose (and other) electrodes is their instability. The stability of the immobilized catalysts is already limited on account of their protein nature, in addition to which the electrode recording is also considerably influenced by other components of the sample solution than those indended for measurement. An example of this is provided by the measurement of glucose via changes in pH: the results are only reproducible if the H^+-concentration and the buffer properties of the samples are identical, or at least known. In addition, adequate oxygen has to be available for the glucose oxidase reaction.

6.4 Enzyme Thermistors

In most enzyme reactions the heat released is in the order of 5 to 100 kJ/mol. Because an enzyme usually reacts only with one specific substance, its substrate, the quantity of heat developed can be utilized for the specific and quantitative measurement of the substrate concerned.

In the so-called enzyme thermistor the temperature change produced when a sample solution is pumped through a column containing immobilized enzymes is measured by a thermistor (sensitive to changes in resistance). By comparing this value with the temperature changes caused by running a solution of known composition through the column, the amount of substance in the unknown sample can be calculated. A typical thermistor arrangement is shown in Fig. 79. The sensitivity depends upon the individual system: the greater the enthalpy of a reaction, the more sensitive is the system. Substrate concentrations as low as a few umol/l are often detectable.

Because the temperature differences arising in enzyme thermistors are extremely small (m°C), it is essential to eliminate interfering factors as far as possible, or at least to keep them constant. Fluctuations in the external temperature can largely be excluded by packing the apparatus in metal, insulating with foam and immersing the entire setup in an ultra-thermostat waterbath. Other sources of interference, such as heat of friction, are more easily overcome in the double column thermistor shown in Fig. 80 than in the single column version. The reference column then contains inactivated enzyme, or only carrier particles.

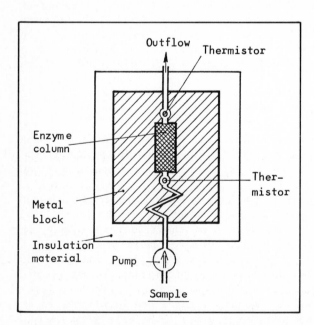

Fig. 79. Enzyme thermistor device with one column

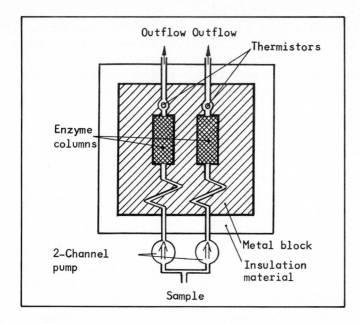

Fig. 80. Enzyme thermistor device with reference column

In principle, it should be possible to use whole cells in connection with thermistors, although the analysis would be much less specific than with single enzymes on account of the large number of side reactions in the whole-cell system.

6.5 Immuno Methods

Antibodies are specific, nonenzymatic proteins formed by humans and animals when high-molecular foreign substances, termed antigens, enter the body. The antigens are rendered harmless by an antibody-antigen reaction which exploits the high affinity of the two substances for one another and results in the formation of antibody-antigen complexes.

Due to the high specificity of this reaction, both in vitro and in vivo, it can be used for diagnostic purposes, to identify either antibody or antigen. Certain immuno methods can be elaborated even further by coupling enzymes to either antibody or antigen. Immuno methods are employed for identifying antibodies and substances that act as antigens, such as high-molecular proteins and carbohydrates.

A number of immuno assays are based on what is known as the ELISA principle (enzyme-linked immuno sorbent assay). To the sample to be analyzed is added a known amount of enzyme-bound antigen, after which the mixture is brought into contact with immobilized antibodies. The enzyme-labeled antigens compete with the native antigens in the sample to form antigen-antibody complexes. Either the bound- or unbound-labeled antigens are then quantitatively determined on the basis of their enzyme labels. From this result it is possible to calculate the quantity of native antigens in the original sample.

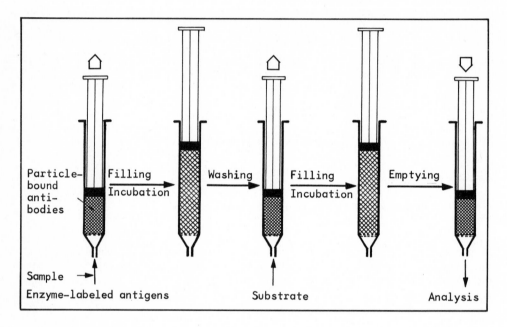

Fig. 81. Immunochemical analysis using enzyme-labeled antigens and particle-bound antibodies

One of the ways of carrying out an immunochemical test of the ELISA type is outlined in Fig. 81. The syringe contains carrier-bound antibody. Following the addition of a known amount of enzyme-labeled antigen to the sample, it is drawn into the syringe and left for a certain length of time in order for antibodies and antigens to come into contact and form complexes. This is followed by washing-out of unbound components, after which a substrate solution is drawn into the syringe. Incubation now leads to enzymatic breakdown of the substrate, measurable by using an appropriate indicator reaction. The greater the amount of substrate converted enzymatically, the smaller is the amount of antigen in the original sample. By comparison with standards a quantitative result can be obtained.

Another system used in immunochemical analysis employs a coupled enzyme, as shown in Fig. 82. The sample to be analyzed is brought into contact with immobilized antibodies. The antigens in the sample form complexes with the carrier-bound antibodies and by means of subsequent washing all other components can be removed. An excess of catalase-coupled antibody is now added and this binds to all of the antigens that are already bound to immobilized antibodies. The larger the amount of antigen in the original sample, the more catalase will be bound. After repeating the washing procedure, H_2O_2 is led through the sample, O_2 is set free in proportion to the quantity of catalase (and thus in proportion to the quantity of antigen in the sample) and can be estimated electrochemically.

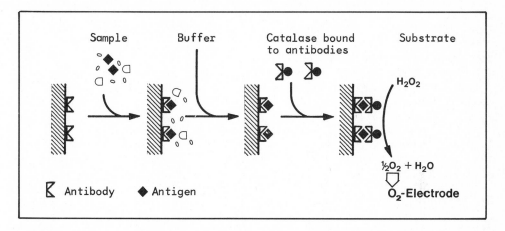

Fig. 82. Example of the determination of antigen using an oxygen electrode and antibody-labeled catalase

Various types of apparatus have been suggested for the kind of system illustrated in Fig. 82. For example, the analysis can be carried out in a flow-through system consisting of a column containing particle-bound antibodies, an oxygen electrode being so situated as to measure the outflow. An automatic analyzer can be constructed on this basic concept by automating the addition of samples and the other solutions required at appropriate points along the system. Another elegant alternative is the direct immobilization of the antibodies on an oxygen electrode which is then automatically brought into successive contact with the sample solution, washing solution, catalase solution, washing solution, and peroxide solution. This contact can be brought about either by immersion of the oxygen electrode in the different solutions or by pumping the solutions over the electrode surface. Such electrodes, if coupled with immunologically active substances and employed with enzymes for analytical purposes, are known as enzyme immunosensors.

7 Uses in Medicine

Immobilized enzymes are already used in medical analysis and diagnosis. The principles of the analytical methods have been dealt with in Chap. 6. In addition, certain of the L-amino acids produced by the methods described in Sects. 5.2 and 5.3 are of therapeutic value, while the methods for producing derivatives of penicillin (see Sect. 5.5) are of the utmost importance in medicine.

The present chapter, gives a brief account of current research into further possible ways in which immobilized enzymes can be put to use in medical therapy.

7.1 Intracorporeal Enzyme Therapy

The administration of enzymes in cases of enzyme deficiency, inborn errors of metabolism and in the treatment of certain types of cancer appears to offer a successful form of therapy. In such cases it is advantageous to use immobilized enzymes because, in human beings as well as in animals, they elicit less antibody production than do the soluble forms of enzymes. Moreover, the enzyme is usually much less likely to be broken down by the body's own proteases when in the immobilized form.

Microencapsulation (in this case it would be more accurate to speak of "nanoencapsulation") is one of the most popular ways of immobilizing enzymes destined for injection. Nanocapsules of this kind can even pass along the finest capillaries of the circulatory system. The capsular material is only permeable to the low-molecular components of the blood and therefore protects the enzyme proteins from contact with proteolytic enzymes, antibodies and antibody-forming components of the blood. Problems have been encountered, however, in connection with elimination of the capsular material.

L-Asparaginase

Some success has been achieved in administering L-asparaginase in capsules made of nylon and polyurea to mice and rats. The enzyme deaminates L-asparagine to L-aspartic acid according to the following reaction:

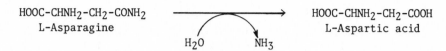

$$\text{HOOC-CHNH}_2\text{-CH}_2\text{-CONH}_2 \longrightarrow \text{HOOC-CHNH}_2\text{-CH}_2\text{-COOH}$$

L-Asparagine   L-Aspartic acid

$$H_2O \qquad NH_3$$

Since L-asparagine is essential for the growth of lymphosarcomas, its enzymatic breakdown within the body offers a promising mode of therapy. As expected, the experiments confirmed that microencapsulation prolongs the activity of the L-asparaginase, to 7 days as compared to the 3 days for which the soluble form remained effective.

Catalase

The enzyme catalase has also been tested in mice. Normally, the blood of mammals contains sufficient amounts of enzyme to break down hydrogen peroxide and any other peroxides that may be present. In a deficiency disease known as acatalasemia, which can occur in mice and men, the catalase is missing.

Experiments aimed at comparing the effects of soluble and encapsulated catalase revealed that the latter form was tolerated by mice even when administered repeatedly, whereas repeated injection of soluble catalase led to death of the animals. The effectiveness of the enzyme injections was tested by injecting sodium perborate. Although the soluble enzyme had a slightly better initial effect, the perborate breakdown achieved in acatalasemic mice was about the same as that of healthy mice, whichever form of enzyme, encapsulated or soluble, was injected.

Perspectives

Investigations in progress are aimed at a better understanding of the breakdown and elimination of the capsular remnants and the enzymes. Great hopes are centered on the encapsulation of enzymes in liquid membranes by the liposome technique. It seems that an appropriate choice of lipids may make it possible to endow the capsules with a specific tendency to accumulate in certain organs and tissues. This would mean that enzymes encapsulated within liposomes could be directed to a specific target, such as lymphosarcoma, for therapeutic ends. With a similar goal in mind, i.e., that of bringing a therapeutic agent to a specific target, experiments are being carried out with enzymes bound to monoclonal antibodies.

7.2 Extracorporeal Enzyme Therapy

Apart from the direct administration of immobilized enzymes, their use in an extracorporeal shunt is being tried experimentally. Shunt systems are in some respects easier and safer than injection; for one thing, they can be switched on and off, whereas, once injected enzymes are not easy to remove or inactivate at will. Figure 83 illustrates the principle of enzyme therapy using immobilized preparations in a shunt. Some degree of success has been registered using L-asparaginase in extracorporeal shunts in test animals (apart from its intracorporeal therapeutic value as described in Sect. 7.1). The enzymes were either immobilized onto poly-methacrylic plates or were microencapsulated. L-Asparaginase bound to the inner wall of nylon tubing has also been used in shunt experiments.

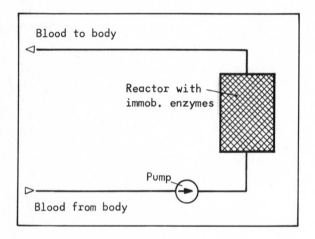

Fig. 83. Extracorporeal enzyme therapy

In addition to substitution therapy in enzyme deficiencies and the treatment of lymphosarcomas already mentioned, the development of shunt systems is of interest with a view to the emergency treatment of acute cases of poisoning. The ideal aimed at would be to have available the appropriate shunt for every possible type of poisoning.

7.3 Artificial Organs

Many attempts have been made to imitate different organs of the body. By far the most attention has been paid to the kidney. Conventional dialysis as practiced on patients with renal disease has the disadvantage of requiring 100 to 300 liters of dialysis fluid per treatment. This quantity could be vastly reduced by pumping the dialysis fluid with its load of urea through an enzyme reactor, as shown in Fig. 84, where the urea would be removed and the dialysis fluid prepared for further use. If kept reasonably compact, a portable artificial kidney could even be constructed on this principle.

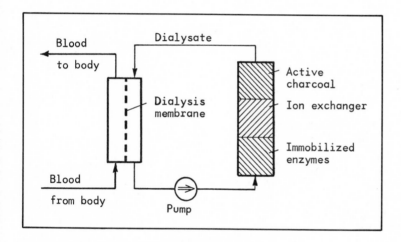

Fig. 84. Scheme of an artificial kidney

The main function that could be taken over by an apparatus of the type shown in Fig. 84, is the catalytic breakdown of urea by immobilized urease, a reaction which proceeds according to the following equation:

$$H_2N-CO-NH_2 \quad + \quad H_2O \quad \xrightarrow{\text{Urease}} \quad CO_2 \quad + \quad 2\ NH_3$$

The ammonia liberated in the reaction is adsorbed onto a suitable ion exchanger, and various other impurities can be removed by adsorption on active charcoal. The carbon dioxide arising in the reaction can be returned to the body in the blood. It is hydrated to H_2CO_3 via the carbonanhydrase system; carbamino hemoglobin is formed and carried to the lungs where the carbon dioxide is finally released and expired in gaseous form.

The development of an artificial kidney is still in its early stages. Although the system shown in Fig. 84 can take over the function of breaking down urea it cannot replace the kidney in removing water or maintaining electrolytic equilibrium.

8 Uses in Basic Research

Immobilized enzymes are better suited for the investigation of some of the problems in basic research than their soluble, native forms. In the majority of the examples described below, the binding of the enzymes to carriers makes it possible to expose them to a succession of different media. With soluble enzymes, however, an extremely laborious procedure is necessary for separating a reaction product from the enzyme protein. Another argument in favor of using immobilized enzymes, either coupled to a carrier or confined within a definite space, is that this is the situation prevailing in the native cell, where many enzymes are membrane-bound or localized in compartments. Such conditions can be imitated by the use of immobilized enzymes, which also have the advantage of being easier to study.

8.1 Structural Studies

The study of the three-dimensional structure of even the most simply constructed enzyme is a very elaborate procedure. A helpful contribution can be made by using carrier-bound enzymes, as shown in Fig. 85, to facilitate the stepwise splitting-off and subsequent study of portions of the amino acid chains.

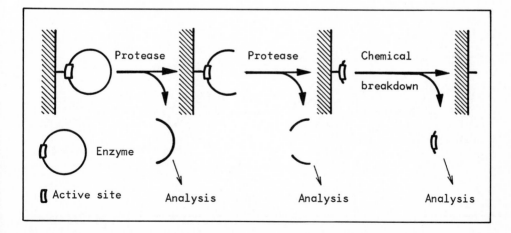

Fig. 85. Sequential proteolytic breakdown in studying the structure and function of a carrier-fixed enzyme

Reversible binding of the enzyme to a carrier via a functional group of the active center (see Fig. 85) makes it possible to begin breakdown of the enzyme protein in the regions more distant from the active center. Simply by washing in buffer, proteolysis can be interrupted at any time and the resulting fragments (in the buffer) studied. Finally, by chemical reversal of the association between enzyme and carrier, the active center and its more immediate vicinity (depending on the extent of proteolysis) can be isolated and analyzed.

8.2 Properties of Enzyme Subunits

Using the immobilization technique important information has been obtained concerning the activity of and the interactions between individual subunits of complex enzymes. Figure 86 shows how a single subunit of a dimeric enzyme can be bound to a carrier. The first step is for the dimer to be bound to a suitable (e.g., bromide-activated) carrier, whose active binding groups are situated so far apart that only one bond can be established between each enzyme molecule and the carrier. If both subunits were to be bound, the object of the procedure would be defeated. The next step is to separate the two monomers by denaturation, after which the nonbound monomer is removed quite simply by washing. Finally, the carrier-bound subunit can be renatured by appropriate treatment.

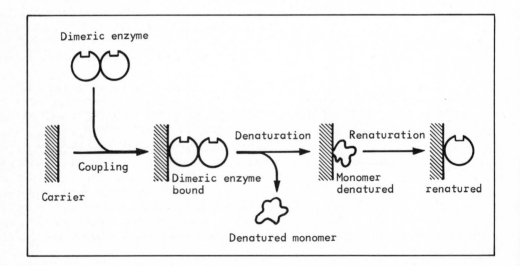

Fig. 86. Carrier-binding of one subunit of a dimeric enzyme

It is easier to investigate the properties of the carrier-bound enzyme subunits (i.e., monomers, see Fig. 86) than to use the soluble form, due to the tendency of monomers from complete dimeric enzymes when they come into contact. This is why it is normally impossible to carry out experiments with soluble monomers.

An immobilized monomer can also be completed to give a normal but carrier-bound dimer, although this kind of procedure is of little practical value since the same goal can be achieved more simply by directly binding the dimer itself to a carrier (see Fig. 86). Of especial interest, however, is the hybridization shown in Fig. 87, in which the catalytic center of the second monomer introduced is blocked, thus resulting in a carrier-bound dimer with one normal and one inactivated catalytic center. The blockage must be stable enough to prevent a reactivation during the subsequent treatment.

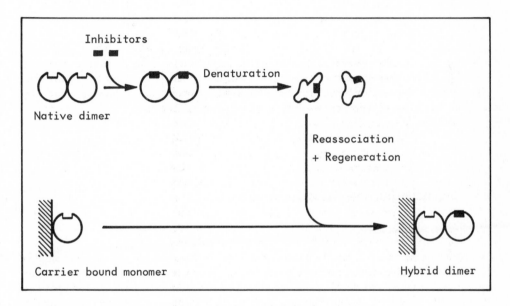

Fig. 87. Formation of a carrier-fixed hybrid enzyme

In comparing the activities of the bound monomer, the normal and the bound dimer, the results may fall into one of the two principal categories, denoted A and B in Table 22. The results shown under heading A indicate that the enzyme is also active when in its monomeric form; the second possibility, B, indicates that the monomer is inactive, but that dimerization leads to a return of activity, showing that the reactivation is a result of interaction between the two subunits.

By starting off with a bound hybrid enzyme and dissociating it from its carrier, it is possible to produce a soluble hybrid enzyme. It is important to compare the kinetic data of the soluble and the immobilized dimers in order to assess the influence of carrier binding.

Table 22. Typical activity values of carrier-fixed monomers, normal dimers
and hybrid dimers

| Carrier-bound form of enzyme | | Relative activity values | |
Form	Symbol	Case A	Case B
Monomer		50 %	0 %
Normal dimer		100 %	100 %
Hybrid dimer		50 %	50 %

The carrier-binding method can also be employed with oligomeric enzymes
(i.e., with more than two subunits) for determining enzyme kinetics,
modulator effects, or interactions between the various subunits. For the
sake of simplicity the few examples discussed here concern only dimers.

8.3 Degeneration and Regeneration

It is often better to use soluble enzymes in studies on denaturation and
regeneration than the carrier-bound form, because the additional carrier
material interferes with many of the modern methods for investigating
structure, such as measurements of fluorescence, or nuclear resonance
studies. In spite of this, immobilized enzymes are sometimes used, usually
on account of the advantage offered by the speed and simplicity with which
the necessary agents can be brought into contact with the bound enzyme and
then withdrawn.

Experiments such as the one illustrated in Fig 88 have resulted in a
better understanding of regeneration. Using covalently bound trypsin it
could be shown that thermal inactivation, generally regarded as being
irreversible, can, in fact, be reversed under certain conditions.

It is generally accepted that trypsin inactivated at 80° C cannot be
reactivated by returning it to optimal conditions. However, cases of
regeneration have been described in which, as Fig. 88 shows, a complete
unfolding of the protein chain by means of urea treatment was intercalated
between the heat inactivation and regeneration. It is also essential for
the SH groups to be protected by the addition of EDTA and cysteine or

other agents serving the same purpose. Heat inactivation apparently causes
an error in the folding of the trypsin chain, making a return to the
active, normal state impossible. Only a complete unfolding of the protein
chain, i.e., an even further departure from the native structure, makes it
possible for the chain to refold correctly to give an active enzyme.

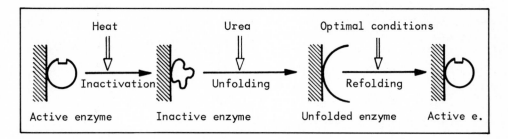

Fig. 88. Regeneration of a thermally inactivated enzyme

Regeneration has in some cases been shown to be promoted by the addition
of a competitive inhibitor. A possible explanation for this is shown in
Fig. 89. The process of refolding allows the possibility of several conformations, correct and false, also perhaps in equilibrium with each
other. If a competitive inhibitor is present it binds only to the "correct" active centers; the correct conformation is now no longer in equilibrium and therefore more of this correctly folded enzyme is formed. It
is easy to imagine that the competitive inhibitor, on account of its
similarity with the true substrate, acts as a kind of "mold" in shaping
the active center.

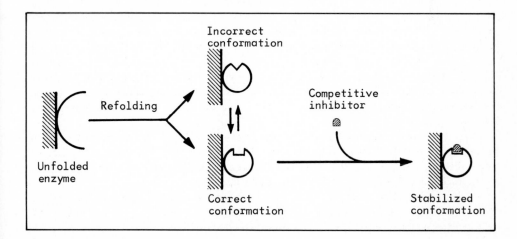

Fig. 89. Regeneration with addition of a competitive inhibitor

Unfortunately, only in a few exceptional cases has it been possible to achieve high yields in regenerations of the type illustrated in Figs. 88 and 89. Much more detailed knowledge must be accumulated before generally applicable, or even practically relevant conclusions can be drawn. Immobilized enzymes will certainly figure increasingly in such studies.

A particularly important role is likely to be played by immobilized enzymes in studying de- and regeneration in oligomeric enzymes. As shown in Sect. 8.2, carrier binding permits the otherwise difficult description and investigation of the individual subunits of enzymes with a quaternary structure.

8.4 Simulation of Natural Systems

Although considerable caution is advisable in extrapolating from "model" immobilized systems to the situation in vivo, experiments with immobilized enzymes may at least permit a closer approximation to the situation in the living cell than can be obtained by in vitro experiments with soluble enzymes. In their natural environment, many intracellular enzymes are membrane-bound or confined by membranes, i.e., are immobilized.

An interesting example of the imitation of natural conditions is shown in Fig. 90. It has been revealed that some enzymes, if attached to nylon via different binding sites, respond to repeated alternate stretching and relaxation of the nylon carrier with an alternating decrease in and recovery of activity. In the case of trypsin, for example, almost 100 % inactivation has been reported to result from stretching the carrier by only a few percent. A plausible explanation, illustrated in Fig. 90, is that stretching deforms the active center. By using immobilized systems of this kind it ought to be possible to arrive at least at an appropriate model for the presumed role of mechanical factors in enzyme regulation in muscle.

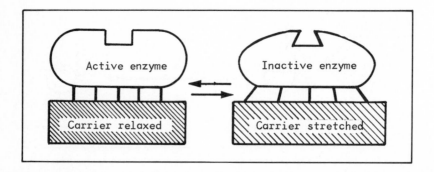

Fig. 90. Enzyme molecule bound to flexible nylon

Another possible use for immobilized enzymes is in the study of multienzyme systems such as those involved in fatty acid synthesis, the respiratory chain or glycolysis. The partial simulation of in vivo conditions
with bound or encapsulated enzymes ought to bring further valuable contri-
butions toward elucidating the diversity of interrelations in such
systems.

9 Special Developments and Trends

What can be termed the almost classical fields of immobilization of single enzymes on the one hand, and whole microorganisms on the other, have recently been extended in a variety of ways to include plant and animal cells and even cell organelles. Although it is too soon to say what significance such advances will have for science and production technology, a few of these special developments and trends will be discussed in the following sections.

9.1 Immobilized Plant Cells

Hitherto, the use of immobilized microorganisms on an industrial scale has invariably served the purpose of obtaining a single enzyme (glucose isomerase, penicillin acylase, aspartase). It is even undesirable in such processes for the microorganism to remain alive. The situation is entirely different with immobilizing plant cells. Most experiments with plant cells are aimed at the synthesis of secondary products, and this requires the participation of complex metabolic reaction chains which, as far as is known at present, can only function when the cell is alive.

Plant cells intended for immobilization are usually obtained by cell culture methods. Figure 91 shows how a plant cell culture is set up. Single cells taken from fragments of the tissue capable of producing the desired end substance are sterilized and explanted onto a solid nutrient. After an interval of several weeks, the cells have multiplied to form a callus, whose constituent cells can either be used for immobilization or allowed to multiply further in a suspension culture.

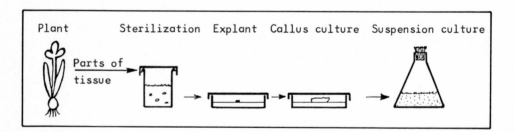

Fig. 91. Procedure for plant cell cultivation

The methods so far employed for immobilizing plant cells have all involved entrapment of the cells in polymer matrices; in the majority of cases natural polymers such as alginate, carrageenan, pectinate, agar, agarose or gelatine have been used. Synthetic materials are usually unsuitable because of the often toxic nature of the monomers used in their preparation and the unphysiological nature of the hardening conditions required. Even the post-hardening of biopolymers with glutaraldehyde for example, leads to cell death and is therefore useless for most of the known fields of application of immobilized plant cells (see Table 23).

The first attempts to use plant cell cultures on an industrial scale have been reported from Japan, where a substance possessing antimicrobial and antiphlogistic properties, Shikonin, is being produced on a scale of several kilograms per batch, using cells of *Lithospermum erythrorhizon*. It remains to be seen whether this fundamental breakthrough will bring other procedures, including the use of immobilized plant cells, in its wake.

Table 23. Possible fields of application for immobilized plant cells

Cell species	Type of reaction	Substrate(s)	Target substance
Catharanthus roseus	De novo synthesis	Sucrose	Ajmalicin
Catharanthus roseus	De novo synthesis	Sucrose	Serpentin
Catharanthus roseus	Synthesis from precursors	Tryptamin, Secologanin	Ajmalicin
Daucus carota	Bioconversion	Gitoxigenin	Periplogenin
Digitalis lanata	Bioconversion	Digitoxin	Digoxin
Morinda citrifolia	De novo synthesis	Sucrose	Anthrachinone

In certain fields of application, as well as in studies on intracellular metabolism, the additional treatment of immobilized cells with plasmolytic substances (nystatin, ether, lysolecithin) is being investigated. Treated in this way, the cell membrane becomes partially or even completely permeable, allowing certain metabolites which normally accumulate within the cell to escape into the surrounding medium. Conversely, other substances which are unable to cross the cell membrane when it is intact can now more easily be introduced into the cell.

Generally, however, cellular metabolism is drastically affected by plasmolysis, so that the cells eventually die and no longer produce the desired substance. A long-term goal is to achieve reversible permeability, which would permit alternating phases of production and harvesting of the product, or, in other words, a kind of intermittent "milking" of the cell.

9.2 Immobilized Mammalian Cells

In the roller bottle and microcarrier techniques for the artificial culture of animal cells the cells are immobilized by their binding to a solid material or carrier. The principal steps in the process by which cells come to adhere to a carrier are shown in a simplified form in Fig. 92. First of all, glycoproteins from the culture medium settle on the surface of the carrier. As soon as a cell comes into contact with the glycoprotein-covered surface it exudes multivalent heparin sulfate at the site of the contact, with the result that the cell sticks more firmly to the carrier. As it adheres, the cell spreads out and clins, in a manner of speaking, to the carrier; in this bound state it is now able to multiply.

Contact with a suitable solid material and spreading of the cells are essential if normal mammalian cells are to multiply when isolated from their normal tissue. This phenomenon is known as anchorage dependence.

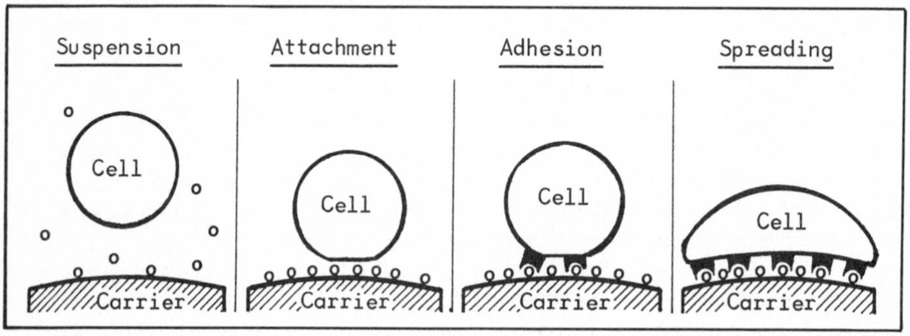

Fig. 92. Adhesion and spreading of mammalian cells on microcarriers
O glycoprotein; ▀ multivalent heparin sulfate

Once attached, the cells overgrow the microcarrier until they form an uninterrupted monolayer. By using microcarriers in simple suspension or in a packed bed it is possible to achieve cell densities of up to several millions per ml. This is 50 times the density obtained by the classical roller bottle method, in which growth proceeds on the inner wall of the slowly rotating bottle.

The microcarrier surface has to fulfill the conditions required for optimal adhesion of the cells, it must be biologically inert, and it must be sufficiently transparent to permit microscopical observation. Size and density of the carrier material have to be such that a good suspension is possible with only a minimum of stirring, and only if these conditions are fulfilled can a high growth rate and cell yield be achieved in a suspension culture.

A variety of synthetics and dextran derivatives can be used as microcarrier materials. The carrier often receives an additional coating of polylysine or collagen. Glass is a good carrier material for use in the packed bed but is less suitable for suspension cultures, since, due to its high density, it requires greater intensity of stirring and this may disturb the cell layer.

Apart from its use in the production of large quantities of cells the microcarrier technique is also employed in metabolic studies, for instance on tumor cells. More than a hundred different cell types can be cultured on microcarriers.

Technical applications of animal cell cultures usually proceed on a scale of between 100 and 1000 liters working volume. The time needed before a culture is ready for use is somewhere between 3 and 8 days.

Table 24. Application of immobilized mammalian cells

Product	Cells cultivated
Polio virus	Human cells from children
Herpes virus	Monkey cells
Mouth and food disease virus	Cells of Syrian hamster
Sindbis virus	Cells of Chinese hamster; chicken fibroblasts
Interferon	Human fibroblasts; skin cells

As Table 24 shows, the chief industrial application is in the large-scale propagation of viruses for vaccination purposes. The initial step in such procedures is the production of large numbers of animal cells. The culture is then infected with viruses which multiply in the cells. After isolation from the culture, the virus particles (usually in the non-viable form) are used in vaccines. A very important recent addition to this field, and one with good prospects for the future, is the production of interferons.

9.3 Immobilized Organelles

Organelles are defined subunits of the cell, surrounded by membranes or systems of membranes, in which certain enzyme sequences are localized. For example, the mitochondria in eucaryotic cells contain the respiratory enzymes and the microsomes contain oxidation enzymes; in green plants the chloroplasts contain the photosynthetic enzymes. Table 25 shows some examples, taken from the literature, of immobilized organelles and the matrices used for their immobilization.

Table 25. Some matrix-entrapped organelles described in literature

Organelles	Origin	Matrix material	Reference
Chloroplasts	Mustard	Polyacrylamide gel	Karube et al. (1979)
Chromatophores	Bacteria	Cross-linked albumin	Garde et al. (1981)
Mitochondria	Yeast cells	Photo cross-linked resins	Tanaka et al. (1980)
Microsomes	Rat liver	Agar	Aizawa et al. (1980)
Peroxisomes	Yeast cells	Different polymers	Tanaka et al. (1978)
Thylakoids	Lettuce	Carrageenan	Cocquempot et al. (1981)

Theoretically, isolated and immobilized organelles could function as excellent biocatalysts in multistep reactions. As compared with the whole cell they would have the advantage that a whole series of undesirable side reactions would be excluded because the enzymes located in organelles are limited to quite definite and specific reaction chains. In addition, the enzymes involved in the reaction chains are present in a high concentration, so that the specific activity of an organelle is high.

Unfortunately, organelles are extremely sensitive and their isolation as well as the subsequent immobilization procedure leads to losses in activity. Relatively speaking, as with plant cells, the best results are obtained by entrapment. The stability of organelles is invariably higher in the immobilized than in the isolated, nonbound state. The stability and productivity of the more complicated organelles, such as chloroplasts or mitochondria, however, is appreciably poorer after immobilization than if they are left in the living cell.

Immobilized chloroplasts and thylakoids have been used in the construction of interesting in vitro photochemical systems for producing energy and reduction equivalents, although industrial application of these and other immobilized organelles is unlikely in the near future.

9.4 Co-immobilization of Enzymes and Cells

The idea behind co-immobilization of enzymes is to supplement the biocatalytic properties of a cell with enzymes that are not present, or only in small amounts, in the cell itself.

Coupling of Enzymes to Dead Cells

The first co-immobilisates that were described in the patent literature in the mid 1970's consisted of enzymes and dead cells. Since the cells usually contained only one enzyme of importance, the processes described can also be regarded as the co-immobilization of two enzymes, one of which was left within the cellular compartment for economic reasons, to save downstream processing.

Figure 93 illustrates a procedure for coupling enzymes to dead bacterial cells. The cells are first treated with a bi- or multifunctional reagent (toluene diisocyanate, epichlorhydrine, glutaraldehyde and others) which kills the cells and provides them with reactive groups. When the enzymes are added, they bind to the free, reactive groups of the bi- or multifunctional reagents which are already bound to the cell wall by their other reactive end. This results in the production of dead cells which have one or more still-active intracellular enzymes as well as additional enzymes coupled to the outside of the cell membrane.

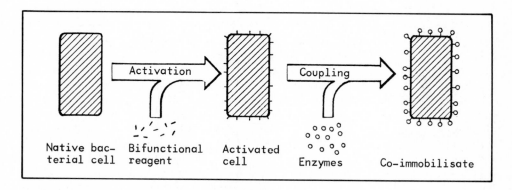

Fig. 93. Enzyme coupling to dead bacterial cells

The method shown in Fig. 93 was designed by Y. Takasaki, mainly for the purpose of coupling glucoamylase to bacterial cells containing glucose isomerase. The biocatalysts thus produced can convert dextrins to glucose-fructose mixtures in a direct sequence in one and the same reactor. The reason why the procedure has not achieved industrial significance is that glucose, which is the usual starting substrate in the production of glucose-fructose mixtures, can be produced both simply and cheaply using soluble glucoamylase. Another serious disadvantage of the co-immobilization system is the poor compatibility of the glucose isomerase and glucoamylase due to the very different pH and temperature required by the two enzymes in order to develop their optimal activity.

Co-entrapment

A method that is universally employed on account of its mildness consists in entrapping the biocatalysts in a common matrix. The method can also be

130

described as co-entrapment, but since the usual entrapping materials are
too coarsely meshed to retain enzymes, special measures have to be adopted.

The example given in Fig. 94 requires preimmobilization of the enzymes
destined for co-entrapment. Immobilization is usually achieved by covalent
binding to carrier materials or by simple cross-linking of the enzymes.
The immobilized enzyme particles are then embedded together with the cells
(as described in Sect. 2.5) in calcium alginate or other natural or synthetic matrices.

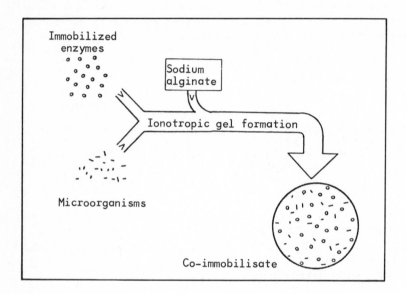

Fig. 94. Co-entrapment of enzymes and living cells

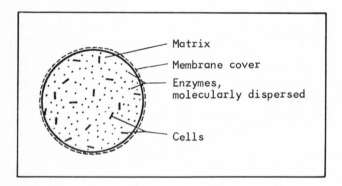

Fig. 95. Membrane-covered alginate bead comprising whole cells and
molecularly dispersed enzymes

In an interesting variant of the co-immobilization procedure, illustrated
in Fig. 94, the enzymes are first of all covalently bound to the still-
soluble matrix molecules (e.g., by the carbodiimide method). The enzyme-
loaded matrix is then used in the normal way for entrapping the living
cell, e.g., by ionotropic gel formation.

Yet another modification of co-encapsulation is shown in Fig. 95. In
this variant, the polymer matrix is surrounded by an additional membrane
which makes possible the retention even of single enzyme molecules. This
is achieved by adding anionic alginate to dispersed cationic acrylate-
methacrylate copolymer, whereby the alginate becomes coated with a film
that fulfills the requirements of a membrane capsule.

Direct Coupling of Enzymes to Living Cells

A disadvantage of the matrix-entrapped biocatalysts is the diffusion
barrier that builds up due to the matrix. This can cause a quite considerable drop in the reaction velocity, especially at low ratios of
actual substrate concentration to K_m values (cf. Sect. 3.6). In the co-
immobilization method shown in Fig. 96, however, no appreciable diffusion
barriers arise. The enzymes are allowed to bind directly to the cell walls
on which they form a thin layer. In the case of robust yeasts, at least,
the cells can be kept alive if they are dried cautiously before being
placed in the enzyme solution where they are bound as follows: on immersion in the enzyme solution the dried cells immediately fill with water,
thereby drawing the enzyme molecules to the cell walls onto which they can
then be deposited by the addition of tannin, cross-linked by glutaraldehyde and bound to the components of the cell wall.

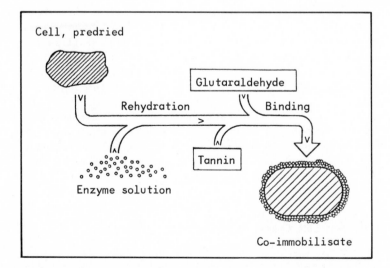

Fig. 96. Direct coupling of enzymes to living cells

In addition to the advantages of a low diffusion barrier, the co-immobilisates shown in Fig. 96 have a disadvantage in that they undergo considerable changes when the cells multiply. When yeast cells coated with enzymes multiply by budding, the new cells thus produced have no covering of enzymes. This can only be prevented by additional precautions such as employing only resting cells or by closely packing the cells in a bed reactor.

Prospects

Most investigations of co-immobilized enzyme-cell systems have been made on fermenting yeasts, whose substrate spectrum was extended by co-immobilization with foreign enzymes. In this way, yeasts of the species *Saccharomyces cerevisiae* were rendered capable of fermenting cellobiose, dextrins and lactose. The reason why co-immobilisates of enzymes and whole cells have not become industrially important is that the systems so far investigated yield only very cheap end products, which can just as well be produced with native biocatalysts. Apart from adding to our scientific knowledge, co-immobilized systems will undoubtedly only play a significant role in isolated cases, for example if the production of a complicated substance (such as an alkaloid or antibiotic) by a living cell can be improved, or even made possible, by the addition of another enzyme.

9.5 Other Co-immobilized Systems

So far, immobilization has followed two main lines of development, i.e., the coupling of single enzymes to inactive carriers and the immobilization of dead or living cells. Recent years have brought the addition of variations and combinations, an example of which, the co-immobilization of enzymes and whole cells, has already been discussed in the previous section.
In a simplified form, Fig. 97 gives an idea of the network of possible combinations of various components. Each of the possible combinations of primary and secondary components in the framework of Fig. 97 has already been described in the literature. Two interesting examples will be discussed below.

Co-immobilization of Different Cell Types

Among the mixed cultures immobilized by allowing them to grow on carrier material are some that have long been used in industrial processes. The best known examples of this, i.e., the production of acetic acid using *Acetobacter* species adhering to wood shavings, and waste water clarification by trickling or percolation over porous material colonized by organisms, have already been discussed in Sect. 5.1. Methane bacteria growing on solid materials such as anthracite are also widely used in a promising technique for producing biogas from certain types of low-turbidity waste water.

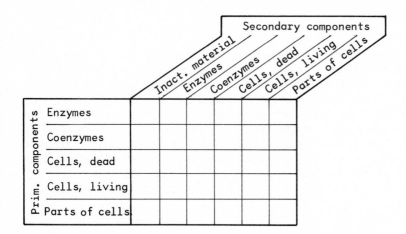

Fig. 97. Major possible components of immobilized biocatalysts

Apart from the immobilized mixed culture systems already in practical use, a number of other possible variations are being investigated, although they will certainly not be put to industrial use in the near future. Co-entrapment of unicellular algae and Clostridia for producing hydrogen from water is one such example; the reaction scheme is shown in Fig. 98.

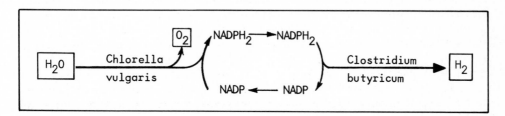

Fig. 98. Scheme of reactions for producing hydrogen by means of a Chlorella-Clostridia co-immobilisate

The unicellular green algae in Fig. 98 bring about photolysis of the water, making available reduction equivalents, chiefly in the form of reduced NADP. The $NADPH_2$ is taken over by the hydrogenase system of the Clostridia and the hydrogen liberated is released as molecular hydrogen.

The reoxidized NADP is then returned to the algae. In the immobilized form, the system delivers appreciably larger quantities of hydrogen than the corresponding free cells because the hydrogenase system of the Clostridia, which is normally highly sensitive to oxygen, is stabilized by immobilization. Nevertheless, the high costs of immobilization and the low productivity make a technical application of the method unlikely.

Co-immobilization of Organelles and Cells

The joint immobilization of organelles and whole cells is also being attempted with the aim of hydrogen production. For this purpose, chloroplasts (e.g. from spinach) are embedded with Clostridium cells (e.g., *Clostridium butyricum*) in agar or other matrices. In its essentials, the hydrogen production follows the same scheme as that shown for the Chlorella-Clostridia system in Fig 98. The Chlorella cells are replaced by chloroplasts, and the NADP, as electron receptor, by ferridoxin or bensylviologen. The main disadvantage of this system is its unsatisfactory half-life, which even in the immobilized setup is only about 10 h.

9.6 Combination of Immobilization with Other Techniques

By combination with other techniques and advances of modern biotechnology, immobilized biocatalysts can be expected to undergo a great boost in popularity. Improvements in our basic understanding of the connections between structure and function of biocatalysts at the molecular level may also provide further impetus, which could, for example, lead to a better understanding of the question of enzyme stability.

Gene Technology

The rapidly developing field of gene technology opens up a whole range of possible areas of application for immobilized biocatalysts.

Using the methods of this new field of technology it is possible to transfer certain enzymes to entirely different organisms belonging to other species. A familiar example, practiced in industry, is the transfer to cells of the bacterium *E. coli* of the ability of human cells to produce insulin. Analogous procedures can be carried out with enzymes; by similar methods they can be transferred from cells that are unsuitable for use in fermentation techniques to cells that are easier to handle (e.g., bacteria or yeasts). Thus, using fermentation technology, it will be possible in the future to produce enzymes that have so far not been available in larger quantities. It is reasonable to expect that among the enzymes and cells obtained from manipulated organisms, some will be immobilized for further application.

Two-phase Systems

Yet another interesting possibility is offered by the use of biocatalysts in two-phase systems. The ideal situation in such a system is given when the substrate and biocatalyst are almost exclusively located in one phase, and the reaction products exhibit a higher affinity for the second phase. In this case, as Fig. 99 shows, the products can be removed continuously; in the frequent cases where they exert an inhibitory effect, their removal would contribute to accelerating the reaction. At the same time fresh substrate can be added, making possible an overall continuous process.

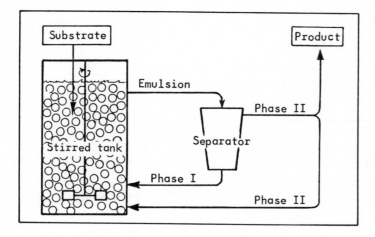

Fig. 99. Procedure for applying immobilized biocatalysts in a two-phase system

Perspectives

We can neither predict how long it will take before the basic knowledge and the practice-orientated developments in the various branches of biotechnology can be put to industrial use, nor can we prophesy the scale of such future undertakings. As we have already seen, a promising beginning has been made in the large-scale use of immobilized enzymes of the second generation. The universally expressed desire for better cooperation between scientists and technicians engaged in the different branches of research justifies the expectation that good progress will be made, not only in the sector of immobilized biocatalysts, but also in the entire field of biotechnology. What form this progress will assume is not within our power to foresee.

PRACTICAL SECTION

Exercise 1 Adsorptive Coupling of Invertase to Active Charcoal

E 1.1 Introduction

This experiment is intended to familiarize the student with the oldest and simplest method of coupling an enzyme to a solid carrier (cf. Sect. 2.1). It will be shown that adsorptive binding is not only easy to bring about, but is equally easily reversed.

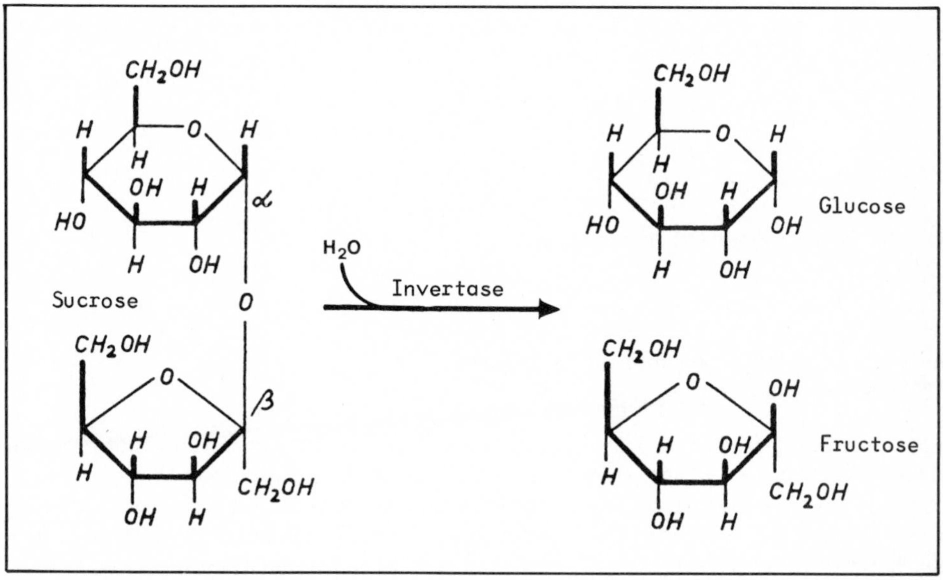

Fig. 100. Enzymatic breakdown of sucrose by means of invertase

Invertase (ß-D-fructofuranoside fructohydrolase, EC 3.2.1.26), as shown in Fig. 100, catalyses the hydrolytic breakdown of sucrose (beet sugar, cane sugar) to glucose and fructose. The resulting mixture of equal amounts of these two sugars rotates the axis of linearly polarized light to the left, i.e., in the opposite direction to the dextrorotary sucrose, and is for this reason also referred to as invert sugar (from the Latin "invertere," to invert). Invert sugar crystallizes out less readily than household sugar, and its use in confectionery thus ensures that the products remain fresh and soft even when kept for longer periods of time.

Soluble invertase is employed in the sweet industry in the production of noncrystallizing creams for the liquefaction, by means of inversion, of fondant fillings for chocolates and for softening marzipan. Invertase is also used in the production of artificial honey, and to a small extent in the industrial production of liquid sugar.

The annual sales of invertase in Europe amount to about 3 million US dollars, although in the USA. it plays a very minor role. The liquid invertase preparations generally employed have an activity of 2400 Sumner units (or 0.8 Weidenhagen units) per ml. On a large scale, invertase is derived solely from yeasts, mostly from *Saccharomyces cerevisiae*. Because invertase is tightly bound to the yeast cell wall, it first has to be exposed by autolysis, or enzymatic breakdown of the cell wall, before it can be further processed to give the final, liquefied market product.

In its immobilized form, invertase has so far only been employed experimentally. Since the soluble enzyme is available at small cost, there is little economic advantage to its repeated and continuous utilization in immobilized form. It remains to be seen, however, whether at a future date the ion exchange procedure for sucrose breakdown, which involves considerable quantities of waste effluent, may not be replaced by methods involving immobilized invertase.

E 1.2 Experimental Procedure

Adsorption of Invertase onto Active Charcoal

An accurately weighed 0.5 g active charcoal is placed into a centrifuge tube, and the amount of invertase solution (Invertase Ingelheim[R], Boehringer Ingelheim KG, D-6507 Ingelheim) indicated in Table 26 is added, plus 0.5 ml 0.2 M acetate buffer (pH 4.5) and 5 ml distilled water. The centrifuge tube is tightly sealed and placed horizontally on a shaker at 25° C for 20 min.

Table 26. Variants for the adsorption of invertase onto active charcoal

Variant	Amount of invertase	Washing with
No. 1	0.2 ml	2x Buffer
No. 2	0.5 ml	2x Buffer
No. 3	1.0 ml	2x Buffer
No. 4	1.5 ml	2x Buffer
No. 5	1.5 ml	1x NaCl + 1x buffer

The active charcoal to which invertase is already coupled is then separated by centrifugation. The resulting residue is stirred up (washed) with diluted buffer (0.5 ml of the above 0.2 M acetate buffer plus 5 ml distilled water) and recentrifuged. The washing procedure is repeated once, the supernatant being discarded each time. To the invertase-containing residue are added 5.5 ml buffer, diluted as above, and the enzyme activity assayed as described below.

In variant No. 5 the residue is washed with 0.5 ml 20 % NaCl solution plus 5 ml buffer instead of the diluted buffer.

Sucrose Breakdown

In 5 ml acetate buffer (0.2 M, pH 4.5) 5 g sucrose are dissolved and made up to 50 ml with distilled water for further use in shaking flasks at 30° C.

For each of the variants 1 to 5, the 50 ml substrate and all of the invertase adsorbed on active charcoal are used. An additional flask is prepared as variant 0, in which adsorbed invertase is replaced by 0.1 ml of the original invertase preparation.

After 0, 5, 10, 15 and 20 minutes an aliquot of approximately 3 ml is withdrawn from each flask and pipetted into a preheated test tube in a boiling water bath to inactivate the invertase. After 2 min the sample is cooled and tested for reducing sugars with dinitrosalicylic acid.

Determination of Reducing Sugars

The dinitrosalicylic acid reagent is prepared as follows: 5 g dinitrosalicylic acid are dissolved by warming with 100 ml 2 M NaOH. Parallel to this, 150 g potassium sodium tartrate are dissolved by warming in roughly 250 ml distilled water. After cooling, the two solutions are combined and made up to exactly 500 ml with distilled water.

For the assay itself, 1 ml of the cooled sample (diluted if necessary) and 2 ml reagent solution are mixed in a test tube, placed in a boiling water bath for exactly 5 min, and then cooled for 3 min in a cold water bath. After dilution 1:10, the extinction is measured through a 1 cm layer at 530 nm against a blank containing the original sucrose solution + buffer + dinitrosalicylic acid reagent, etc. By means of a calibration curve it is possible to read off the amount of reducing sugars (= sum of glucose and fructose) that corresponds to the extinction value of the sample.

If the sample is found to contain more than 3 mg reducing sugar per ml the determination has to be repeated with a more dilute sample.

The amount of reducing sugar in 100 ml is obtained by multiplying the value taken from the calibration curve by 100 and, if necessary, by the dilution factor of the sample. For example, if 1 part sample was diluted with 2 parts buffer (= 1 in 3) the dilution factor is 3.

E 1.3 Results and Evaluation

Values Recorded

The values in Table 27 are typical results obtained from experiments carried out in practical courses. The values are highly dependent on the experimental conditions and the materials used (intensity of washing, invertase, active charcoal).

Table 27. Results from practical courses

		Number of the variant					
		0	1	2	3	4	5
Amount of invertase	ml	0.1	0.2	0.5	1.0	1.5	1.5
NaCl-washing		no	no	no	no	no	yes
Reducing sugars	mg/ml						
after 0 min		1.9	0.7	1.2	1.6	1.8	0.0
after 5 min		7.8	2.1	5.4	4.0	8.0	0.0
after 10 min		13.4	2.7	8.0	9.2	15.2	0.2
after 15 min		20.6	7.5	12.4	15.2	23.1	0.5
after 20 min		24.6	8.3	16.4	20.0	26.9	0.4
after 25 min		29.4	10.8	20.3	24.8	29.2	0.7
after 30 min		33.4	12.0	24.0	28.8	33.0	0.8

From Table 27 it can be seen that scarcely any additional reducing sugar was set free in the breakdown test of variant No. 5. In this variant, in contrast to the otherwise identical No. 4, 20 % NaCl was used for washing: this apparently led to almost total desorption of the adsorbed invertase. It can also be seen that with increasing amount of invertase added to the charcoal (variants 1-4), the amount of enzyme adsorbed also increases (although less than proportionally). This will be discussed below.

Further Evaluation

A graphical representation, as shown for the variants 0, 1, 2 and 3 (Fig. 101), has the advantage that irregularities in the increase in reducing sugars can be detected more easily than in Table 27. The slope of the regression curves in the figure gives the rate of increase in reducing sugars, thus providing a measure of the enzyme activity. These values are presented in Table 28.

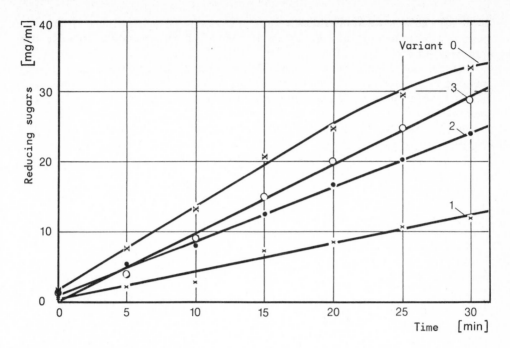

Fig. 101. Reducing sugars as a function of application time

Table 28. Evaluation of activity

Variant No.	Increase of red. sugars mg/ml·min	Initial amount of invertase ml	Increase of red. sugars per invertase used mg/ml·min·ml	Relative activity %
0	1.153	0.1	11.530	100.0
1	0.413	0.2	2.065	17.9
2	0.760	0.5	1.520	13.2
3	0.986	1.0	0.986	8.6
4	1.270	1.5	0.847	7.3
5	0.026	1.5	0.017	0.1

The figures in the last vertical column of Table 28 show that the relative activity in the various assays (as compared with the initial activity presented for binding) drops with increasing initial amounts of enzyme activity. In absolute terms, however, as the last but one column shows, the more invertase present, the more is bound to the active charcoal. In agreement with the general trend in adsorption procedures, invertase adsorption also tends toward a saturation level, above which no more enzyme is adsorbed. In the experiment shown here, a maximum of 17.9 %

binding was achieved. This is a very unsatisfactory binding efficiency, possibly because (to save time) an extremely short period of only 20 min was allowed for adsorption.

It can be assumed (although not investigated in this case) that invertase is not inactivated by the binding procedure since it is known from other experiments that the nonbound activity can be recovered from the washings.

Exercise 2   Ionic Binding of Catalase to CM-Cellulose

E 2.1 Introduction

In this experiment the student will be introduced to the ionic binding of an enzyme to a carrier (cf. Sect. 2.2). It will be shown that the coupling of the enzyme to one of the commercially available ion exchangers is a simple procedure, and that with catalase a high degree of efficiency can be achieved. It will also be shown that bonds between the enzyme and its carrier can be ruptured by other ions.

Immobilization of an enzyme by formation of ionic bonds was first reported by Mitz (1956). He immobilized catalase by running a solution of the enzyme through an ion-exchange column packed with cellulose. Hydrogen peroxide was then pumped through the column and no peroxide was detectable in the effluent.

The simplicity of the procedure by which enzymes can be coupled to cation exchangers is responsible for the fact in 1969 ionically-bound L-aminoacylase became the first immobilized enzyme to be put to industrial use (cf. Sect. 5.2). Despite the relative weakness of the ionic bonds as compared with covalent bonding, enzymes coupled to ion exchangers can be used continuously for weeks and even months, due to the fact that reaction conditions can be kept very constant, particularly the pH value and ionic concentration.

Reversal of the coupling between enzyme and carrier by foreign ions, often a disadvantage, can also be turned to good use: for example by treating with concentrated NaCl solution, partially inactivated enzymes can be completely separated from the carrier and replaced by active enzymes. Since the costly ion-exchange materials can thus be used over and over again, this contributes substantially toward making procedures involving enzymes coupled to ion exchangers economically attractive.

In the present experiment the cation exchanger carboxymethyl cellulose (CM-cellulose) replaces the anion exchanger used by Mitz. Since the isoelectric point (IEP) of catalase is relatively high (about pH 7), the enzyme is mainly present in the protonized (i.e., cationic) form at the pH of 4.5 chosen for our experiment. Under such conditions, coupling to a cation exchanger is much more efficient than using an anion exchanger.

E 2.2 Experimental Procedure

Coupling of Catalase to CM-Cellulose

For each of the variants 1-8 in Table 29, 0.5 g CM-cellulose (Servacel Type CM 23, Serva Feinbiochemica, D-6900 Heidelberg, FRG) are allowed to swell in 6 ml distilled water in a screw-top centrifuge tube, after which 0.1 ml of 0.1 M citrate-phosphate buffer (pH 4.5) and the different amounts of catalase (mold catalase L 43, John & E. Sturge Ltd, Selby, GB) shown in Table 29 are added. The samples are shaken for 60 min at 25° C and then centrifuged. The residue of variants 1-4 is washed with 10 ml diluted buffer solution (0.1 ml of 0.1 M citrate-phosphate buffer in 10 ml distilled water) and centrifuged. This procedure is repeated once more. Variants 5-8 are washed (see Table 29) once with 20 % NaCl solution and once with diluted buffer. The washed residue (CM-cellulose with ionically bound catalase) is used for the assay procedure as described below.

Table 29. Variants of the coupling procedure

Variant	Amount of catalase	Washing with
Nr. 1	0.5 ml	2x Buffer
Nr. 2	1.0 ml	2x Buffer
Nr. 3	1.5 ml	2x Buffer
Nr. 4	2.0 ml	2x Buffer
Nr. 5	0.5 ml	1x NaCl + 1x Buffer
Nr. 6	1.0 ml	1x NaCl + 1x Buffer
Nr. 7	1.5 ml	1x NaCl + 1x Buffer
Nr. 8	2.0 ml	1x NaCl + 1x Buffer

Assay Procedure

0.1 ml of a 30 % solution of hydrogen peroxide and 1 ml 0.1 M citrate-phosphate buffer (pH 4.5) are made up to 100 ml with distilled water and stirred at 25° C. The reaction begins when the catalase-containing material is added. A stopwatch is then started. Per individual assay, only two fifths of the immobilized enzyme preparation is used: this is done by suspending it in diluted buffer (see above) to a volume of 25 ml and using 10 ml of this suspension. The time taken for the peroxide content to drop to 1 mg/ml is recorded. For this purpose, the peroxide content is tested at convenient intervals (e.g., every 30 s) with peroxide test strips (Merckoquant[R] peroxide test, E. Merck, D-6100 Darmstadt, FRG).

Three other variants, i.e., 9, 10 and 11, are run as standards of comparison. In addition to the peroxide solution they contain 0.05, 0.1 and 0.2 ml native catalase, respectively. Each is made up to 10 ml with diluted buffer (see above).

E 2.3 Results and Evaluation

The information that can be drawn from the results in Table 30 is restricted to a certain extent by the fact that the peroxide was estimated in a very simple, semiquantitative way, using peroxide test strips instead of a more accurate, time-consuming method. Furthermore, any measurements in connection with catalase are invariably associated with the fundamental problem that the enzyme is inactivated by its own substrate, hydrogen peroxide. However, the results give us an idea of some of the major tendencies that are of general validity in the binding of enzymes via electrostatic forces.

Table 30. Data measured and derived

Var. No. –	Amount of catalase in immobil. ml/0.5g carr.	in test ml	Time min	Rec. time min^{-1}	Catalase activity per ml cat. $min^{-1} \cdot ml^{-1}$	per g carr. $min^{-1} \cdot g^{-1}$	Relative yield %
1	0.5	0.2	5.5	0.182	0.910	0.91	76.5
2	1.0	0.4	5.3	0.189	0.473	0.95	39.7
3	1.5	0.6	4.5	0.222	0.370	1.11	31.1
4	2.0	0.8	4.0	0.250	0.313	1.25	26.3
5	0.5	0.2	8.5	0.118	0.590	0.59	49.5
6	1.0	0.4	7.5	0.133	0.333	0.67	28.0
7	1.5	0.6	7.0	0.143	0.238	0.72	20.0
8	2.0	0.8	6.0	0.167	0.209	0.84	17.6
9	–	0.05	17.5	0.057	1.140	–	
10	–	0.1	8.5	0.118	1.180	–	100.0
11	–	0.2	4.0	0.250	1.250	–	

The calculated relative yields shown in Table 30 are based on the average of the results from variants 9-11, i.e., $1.19\ min^{-1} \cdot ml^{-1}$, which was taken as 100 %.

In the individual assays, the time taken for the initial peroxide concentration of 300 mg/ml to drop to 1 mg/ml was measured (see column 4 of Table 30): the reciprocal values (column 5) therefore provide a measure of the amount of peroxide broken down per time. The relative specific activity per ml of the catalase initially added can be calculated by relating the amount broken down to the amount of catalase (column 5) or of carrier (column 6) used.

Table 30 reveals that the preparations washed with NaCl (variants 5-8) had a much lower activity than those that were only washed with buffer (variants 1-4), although the loss of activity is less than in Exercise 1 (adsorptive binding).

As was also seen in connection with adsorptive binding (cf. Exercise 1), the amount of enzyme ionically bound to the carrier also increases with increasing amounts of enzyme used, up to a certain saturation point (see Fig. 102). The ability of the carrier to take up more enzyme is exhausted as soon as its surface is completely covered by a monomolecular layer of enzymes. In contrast to the specific activity, the relative yield of carrier-bound enzyme per quantity of enzyme added drops as the amount of added enzyme increases.

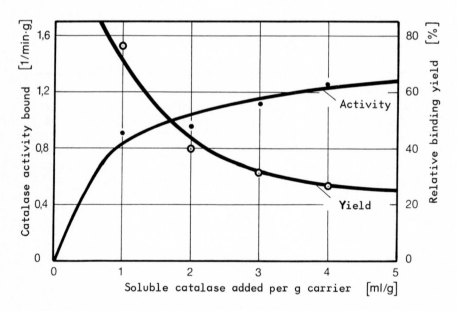

Fig. 102. Yield and activity of bound catalase as a function of the catalase added per carrier

The results show that by using very simple methods, high values of enzyme activity can be coupled to carrier materials by means of electrostatic interactions. Whether the technological development of such methods is economically feasible depends to a large extent upon whether interfering ions and fluctuations in concentration can be eliminated.

Since the unbound enzymes usually remain active, the relatively low yield in relation to the amount of soluble enzyme employed is no great disadvantage in industrial applications. In practice, the ion-exchange material is loaded with enzymes either by stirring the exchanger in the enzyme solution or by pumping a solution of the enzyme through a column filled with the carrier. In both cases the unbound enzymes left in the aqueous solution can be reused or recirculated through the ion-exchange column until almost 100 % coupling is achieved.

Exercise 3   Covalent Binding of Glucoamylase to a Carrier bearing Oxirane Groups

E 3.1 Introduction

The very important method of coupling single enzymes to carriers by covalent bonds will be demonstrated in this exercise by coupling glucoamylase
to an activated carrier. At the same time the experiment serves to
acquaint the student with the enzyme glucoamylase (= amyloglucosidase)
which, in its soluble form, is much used in industrial processes.

Glucoamylase splits glucose units off the nonreducing end of the
chains of amylose, amylopectins and other α-glucans by hydrolyzing the
α-1,4 linkages between the glucose molecules of the oligomers and polymers. Glucoamylase also splits α-1,6 linkages in branched chain glucans
such as amylopectin, albeit at a very low reaction rate.

On account of its ability to split off glucose from glucose polymers,
the enzyme glucoamylase is very widely employed in the starch industry for
obtaining glucose and other glucose-containing products. In distilleries
in which starch-containing raw materials are processed, glucoamylase is
used for sweetening mashes dextrinized by heat and α-amylase.

Numerous ways of immobilizing glucoamylase have been tried out, such
as coupling to various carrier materials, cross-linking, and application
in membrane reactors, and in some cases even developed on a semi-
industrial scale. In spite of such efforts, immobilized glucoamylase is
still not used to any significant extent in industrial processes. The main
reasons for this are probably the low cost of soluble glucoamylase and the
simplicity of the technique in which it is employed. It is, in fact, one
of the cheapest of all enzymes and is produced as an extracellular enzyme
by molds, particularly of the species *Aspergillus* and *Rhizopus*.

Of the multitude of methods available for the covalent binding of
glucoamylase to a carrier we have chosen as an example its coupling to a
carrier bearing reactive oxirane groups (cf. Sect. 2.3). Like the cyanogen
bromide-activated carriers, this type of carrier is also commercially
available. Due to their high costs, such carriers are only used on a
larger scale with expensive enzymes, and therefore for the production of
substances of a much higher price-class than glucose.

E 3.2 Experimental Procedure

Coupling

In 5 ml potassium phosphate buffer (pH 6) 50 mg glucoamylase (Gluczyme[R] 8000, Amano Pharmaceuticals, Nagoya, Japan) are dissolved in each of a series of screw-top centrifuge tubes to which the appropriate amount of carrier (Eupergit[R] C, Röhm GmbH, D-6200 Darmstadt, FRG) is then added. The tubes are carefully closed and left to shake in a horizontal position for 24 h at 25° C. Following sedimentation and careful removal of the super-natant (to be saved!) with a Pasteur pipette, the residue (= immobilized preparation) in variants 1-4 is washed five times more with 0.1 M citrate buffer (pH 4.5)

All supernatants from the washing of one variant are combined and made up to 50 ml with 0.1 M citrate buffer (pH 4.5). These supernatants are tested for activity in just the same way as the residues. The residues (= immobilized preparations) of each variant are suspended in 0.1 M citrate buffer (pH 4.5) to give an exact volume of 25 ml.

Table 31. Variants of the covalent binding of glucoamylase to Eupergit[R]

Variant	Amount of glucoamylase	Amount of Eupergit	Ratio enzyme:carrier
No. 1	0.050 g	0.25 g	0.200 g/g
No. 2	0.050 g	0.50 g	0.100 g/g
No. 3	0.050 g	0.75 g	0.067 g/g
No. 4	0.050 g	1.00 g	0.050 g/g

Assays

In 0.1 M citrate buffer (pH 4.5) 50 ml of a 2.5 % solution of maltodextrin (Snowflake[R] maltodextrin Type 01911, Maizena GmbH, D-2000 Hamburg, FRG) are warmed at 40° C on a shaker, after which 5 ml of the solution or suspension to be assayed are added. In a parallel blank, 5 ml distilled water are added instead of the enzyme-containing sample. The assay tubes are then shaken at 40° C for exactly 5 min. To inactivate the enzyme, about 2 ml are quickly transferred to a prewarmed test tube in a boiling water bath.

The glucose content of each heat-inactivated sample and of the blank is determined by the hexokinase method (glucose test combination No. 716.251, Boehringer Mannheim GmbH, D-6800 Mannheim, FRG).

To determine the activity of the soluble enzyme, 20 mg glucoamylase are dissolved and made up to 50 ml in 0.1 M citrate buffer (pH 4.5) for assay as variant No. 5. This is the standard of comparison for judging the yields and activities of the four variants.

E 3.3 Results and Evaluation

The measured and calculated results of a typical test series are shown in Table 32.

The sum of the activities in the immobilisate and supernatant of the individual variants amounts to between 9.533 and 11.321 μmol/g·min (units/g). This is between 65 % and 70 % of the initial soluble enzyme activity, which was 14.758 units/g. The loss of activity may be the result of enzyme inactivation during the immobilization procedure, which is very common in covalent binding methods. Another possible explanation for the lower efficiency of the glucoamylase after immobilization is that the active centers of the coupled enzyme may, for steric reasons, be less easily accessible to the relatively large molecules of the substrate, dextrin.

Table 32. Results from trials with soluble Eupergit-bound glucoamylase.

Variant No.		Initial enzyme g/55ml	Glucose concentration reference g/l	main value g/l	Increase in glucose μmol/55ml·5min	Activity per enzyme μmol/min·g
1	Immobilisate	0.010	0.381	0.803	128.9	2.579
1	Supernatant	0.005	0.381	1.045	202.8	8.116
2	Immobilisate	0.010	0.381	0.941	171.1	3.422
2	Supernatant	0.005	0.381	0.881	152.8	6.111
3	Immobilisate	0.010	0.381	1.244	263.7	5.274
3	Supernatant	0.005	0.381	0.864	147.6	5.903
4	Immobilisate	0.010	0.381	1.313	281.9	5.638
4	Supernatant	0.005	0.381	0.846	142.1	5.683
5	Native enzyme	0.002	0.381	0.864	147.6	14.758

Table 32 shows that the coupling efficiency (i.e., the amount of coupled enzyme as a percentage of the initial enzyme activity added) increases with an increase in the proportion of carrier to enzyme. At 0.2 g enzyme per 1 g carrier, the coupling efficiency was 17 %, whereas with 0.05 g enzyme per g carrier the efficiency was 68 %. Although the quantity of enzyme bound to 1 g of carrier increases with increasing amount of available carrier, any additional enzyme activity over and above that already bound, is less efficiently bound. This is to be expected, since the binding sites of the carrier are already enzyme-saturated. In the case in question, this saturation point was probably not yet reached with approximately 600 enzyme units per g carrier. The number and variety of tests carried out are sufficient to indicate general trends, but give no accurate information on the interrelationship between the different factors involved.

Exercise 4 Immobilization of ß-Galactosidase by Cross-linking

E 4.1 Introduction

This exercise demonstrates that, using the cross-linking method discussed in Sect. 2.4 it is possible to obtain immobilized preparations with a high activity toward low-molecular substrates. In addition, the student is introduced to the enzyme ß-galactosidase (= lactase), for which a growing role in industrial processes is predicted (cf. Sect. 5.7).

For use on an industrial scale, ß-galactosidases are usually obtained from yeasts, especially *Kluyveromyces marxianus*, or from molds, chiefly *Aspergillus niger* and *Aspergillus oryzae*. Soluble ß-galactosidase, at prices around 100 US$/kg, are among the most expensive of technical enzymes. For this reason, there is much to recommend their immobilization, with the resulting advantages of repeated and continuous use. In a number of firms, the use of immobilized lactases is currently at the experimental stage.

Cross-linking, like the covalent binding of enzymes to carriers, is a relatively harsh method and can lead to considerable changes in conformation and losses in activity. If glutaraldehyde is used as linking reagent, coupling takes place mainly via the free ε-amino groups of lysine. If lysine is present at or in the immediate vicinity of the active center of the enzyme, the reaction with glutaraldehyde leads to inactivation.

Yeast ß-galactosidases are among the relatively few enzymes which, for the above reason, i.e., binding of their essential lysine residues, are not suitable objects for cross-linking with glutaraldehyde. Instead, such ß-galactosidases can successfully be spun into cellulose acetate fibers, a procedure that is also used industrially (see Exercise 8).

ß-Galactosidase from molds, on the other hand, can successfully be immobilized by cross-linking with glutaraldehyde, as the present exercise will show. In one of the variants of the experimental procedure to be described below, egg albumin is bound in the preparation, in addition to the enzyme molecules. This leads to a thinning out of the immobilisates, or, in other words, the enzyme molecules are not so crowded and it is therefore likely that more activity will be retained after cross-linking.

E 4.2 Experimental Procedure

Cross-linking

In 15 ml distilled water 600 mg ß-galactosidase from *Aspergillus oryzae* (Lactase F, Amano Pharmaceuticals, Nagoya, Japan) are dissolved in a centrifuge tube which is then placed in an ice bath. Now 30 ml ice-cold acetone are now slowly added under continuous stirring, followed by 2 ml 25 % glutaraldehyde solution. The sample is then shaken for 60 min at 30° C. After centrifugation, the supernatant is discarded and the residue stirred up with 40 ml distilled water and homogenized with an Ultra-Turrax. Centrifugation is repeated, the supernatant discarded, and the residue washed again with 40 ml distilled water. Finally, the entire cross-linked preparation is suspended to a volume of 100 ml, with the addition of a few drops of sodium azide to prevent microbial contamination. This preparation is tested for activity in variant No. 1.

Co-cross-linking

In principle, the procedure is the same as that for normal cross-linking, except that 150 mg albumin (egg albumin, order No. 18801, Riedel de Haen, D-3016 Seelze, FRG) are dissolved with the 600 mg ß-galactosidase. The resulting preparation is tested for activity as variant No. 2.

Activity Assay

The determination of activity using ONPG (o-nitrophenyl-ß-D-galacto-pyranoside) is based on the reaction shown in Fig. 103, in which the ß-galactosidase releases yellow nitrophenol from the colorless ONPG. The reaction can be quantified by means of spectrophotometry. ß-Galactosidases have a higher affinity for ONPG than for lactose, as can be seen from the lower K_m value.

In 0.1 M acetate buffer (pH 4.5) 370 mg ONPG are dissolved and made up to 100 ml. Of this substrate solution, 4 ml are warmed in a small Erlenmeyer flask in a shaking water bath at 37° C. Then 1 ml of the diluted enzyme solution (or suspension) is then added and a timer is started. After exactly 15 min incubation in the shaker, an aliquot of 1 ml is pipetted into a test tube already containing 1 ml 10 % Na_2CO_3 solution. After rapid mixing, 8 ml distilled water are added. The extinction of this solution is now measured at 420 nm at a layer depth of 1 cm against a blank containing previously inactivated enzyme solution (or suspension). If the difference between the extinction values of sample and blank is outside the range of 0.1 and 0.5 the assays should be repeated with a more diluted or concentrated sample. As a rule, the initial 100 ml containing the cross-linked enzyme particles in suspension have to be diluted further, 1 to 200, and 1 ml of this dilution used for activity assay. This means that of the 600 mg enzyme, 0.030 mg (0.00003 g) should be present in each assay.

An assay with the original soluble enzyme is also carried out as variant no. 3. In 100 ml buffer 60 mg ß-galactosidase are dissolved for this purpose. This solution is further diluted 1:50, and 1 ml of this final dilution is used for the activity assay, i.e., 0.012 mg (0.000012 g) enzyme are present in this sample.

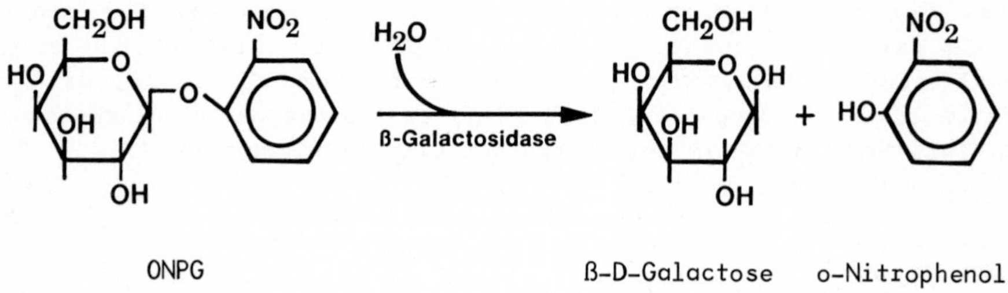

Fig. 103. Breakdown of ONPG by ß-galactosidase

The activity is expressed in lactase units (LU), whereby a lactase unit is defined as the amount of ß-galactosidase that releases 1 μmol nitrophenol per min under the test conditions (37° C and pH 4.5). The activity is given by the relationship

$$\text{Activity in LU/g} = \frac{\triangle E \times 50}{\varepsilon \times 15 \times W}$$

in which $\triangle E$ is the extinction difference between test and blank samples, ε the extinction coefficient and W the amount of enzyme, in grams, contained in the 1 ml of solution added to the ONPG reaction flask. The extinction coefficient is approximately 4.7, but this should be checked using o-nitrophenol as test substance, and redetermined if necessary.

E 4.3 Results and Evaluation

The ß-galactosidase was cross-linked with a yield of about 40 % of the initial activity added. The results in Table 33 indicate that the co-coupling of lysine-containing albumin resulted in only a slightly higher activity value than that obtained by cross-linking the enzyme alone. The high yield of coupled product achieved in this exercise is not unusual for cross-linked enzymes with a low-molecular substrate. But this high yield alone is not usually a sufficient reason for employing such preparations on an industrial scale. A crucial problem is that the gelatinous nature of the immobilized particles renders them unsuitable for use in thicker layers in a packed bed reactor.

Table 33. Data from measurements and calculations

Var. No.	Form of ß-galac-tosidase added	Amount of enzyme in test	Difference in extinction	Activity[a]	Binding yield[a]
1	Cross-linked without albumin	0.000030 g	0.178	4,632 LU/g	35.6 %
2	Co-cross-linked with albumin	0.000030 g	0.204	5,308 LU/g	40.8 %
3	Native (soluble) enzyme	0.000012 g	0.200	13,010 LU/g	–

[a] Activity and binding yield are given in relation to the part of the enzyme used in the cross-linking batch

Exercise 5 Alginate Entrapment of Yeast Cells and Their Co-entrapment with Immobilized ß-Galactosidase

E 5.1 Introduction

This exercise is intended as an introduction to entrapment in polymer matrices, a technique widely used for immobilizing whole cells. In addition, in order to enhance their substrate spectrum, cells will be co-entrapped with an immobilized enzyme.

The basic difference between the coupling of biocatalysts either to one another or to solid carriers, as practiced in Exercises 1-4, and entrapment methods, is that in the latter, normally no coupling reaction takes place between matrix and biocatalyst. This means that such methods are generally very mild, so that they are especially suitable for the immobilization of living organisms. On account of their large pore diameter, most matrices are unsuitable for encapsulating molecularly dispersed enzymes. This disadvantage can be overcome by cross-linking the enzymes before entrapment, as in variant 2 of this exercise.

Of the matrices available, those of biological origin are particularly widely employed; a typical example is the alginate used in this experiment. Alginate is a polyuronide (molecular weight 12,000-120,000) obtained from brown algae, and has the advantages of being harmless even for sensitive cells and of being convertible from the soluble to the insoluble form by a very simple process of ionotropic gel formation. The alginic acid consists of units of ß-D-mannuronic acid and α-D-guluronic acid, linked by α-1,4-glycosidic bonds. Its sodium salt is soluble in water, but when added to a solution of calcium chloride the sodium ions are replaced by calcium ions and the alginate gelates. The higher the proportion of guluronic acid in an alginate, the harder is the ionotropically formed gel.

The species of yeast used in the following experiments, *Saccharomyces cerevisiae*, has for thousands of years played a dominant role in fermentation processes, as wine-, beer-, distiller's-, and baker's yeast. The immobilization of yeast and its application in this form, although often studied, have not acquired any great significance because the production of native yeast in the required quantities is both easy and cheap. Being readily available in large quantities, yeast is particularly well suited for demonstrating the possibilities opened up by immobilization. Furthermore, its metabolism has been thoroughly elucidated and its most important metabolic activity, the production of ethanol, is easy to demonstrate and can be accurately measured.

In one of the variants of this exercise, yeast and immobilized ß-galactosidase of mold origin will be entrapped in a common matrix. This procedure is aimed at endowing the yeast with the additional capacity to split, and thus to ferment lactose.

E 5.2 Experimental Procedure

Entrapment

In 35 ml distilled water 1 g sodium alginate (Manugel^R DJX, Alginate Industrie GmbH, D-200 Hamburg 11, FRG) is dissolved and thoroughly stirred with 10 g fresh baker's yeast. Using a narrow-tube funnel or a hypodermic syringe (without needle) the mixture is then dropped into 200 ml of 2 % $CaCl_2$ solution which must be continuously, but not too vigorously, stirred during this addition. To achieve sufficient hardening of the alginate beads they are left for 1 h in the $CaCl_2$ solution and then stored in 0.5 % $CaCl_2$ solution in the refrigerator until needed.

Care should be taken over this quantitative step, i.e., no trace of the yeast-alginate should be left in the dropping device. Per fermentation experiment exactly half of the alginate beads (corresponding to an initial 5 g of yeast) are used.

Co-entrapment

Half of the ß-galactosidase cross-linked with albumin in Exercise 4 is washed thoroughly several times with distilled water to remove all traces of sodium azide. The washed residue is mixed to a homogeneous mass with 1 g sodium alginate, 10 g fresh baker's yeast and 35 ml distilled water. This mixture is then allowed to drop into 200 ml 2 % $CaCl_2$ solution (as described above) under continuous stirring. The alginate beads are left to harden for 1 h in 2 % $CaCl_2$ solution before storage in the refrigerator in 0.5 % $CaCl_2$ solution.

Fermentation Experiments

Using the apparatus set up as shown in Fig. 104, the alginate beads and fresh yeast are tested for fermentation activity toward glucose and lactose substrates.

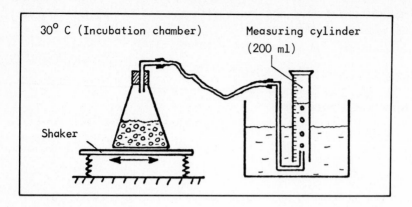

Fig. 104. Experimental setup for determining fermentation activity

Each fermentation experiment is carried out with half of the total immobilized or co-immobilized preparation (corresponding to a starting quantity of 5 g yeast) in 50 ml of an 8 % solution of glucose or lactose in 0.5 % $CaCl_2$. Two additional experiments are carried out, one with 5 g yeast and glucose substrate, and one with 5 g yeast and lactose. A synopsis of the various experiments is shown in Table 34. The quantity of CO_2 set free is read off and noted every 15 min for 4 h. To take readings, the measuring cylinder is pushed into the water bath until the water level is the same inside and outside the cylinder.

Table 34. Variants of the fermentation trials

Var. No.	Substrate	Biocatalyst(s)
1 G	Glucose	Yeast, entrapped
1 L	Lactose	Yeast, entrapped
2 G	Glucose	Yeast + ß-galactosidase, co-entrapped
2 L	Lactose	Yeast + ß-galactosidase, co-entrapped
3 G	Glucose	Yeast, native
3 L	Lactose	Yeast, native

E 5.3 Results and Evaluation

As was to be expected, neither in the native form nor entrapped in alginate was the yeast capable of fermenting lactose, whereas in either form it was able to ferment glucose at greater or lesser speeds. The values (from an experiment carried out in a practical course) also clearly show that the yeast entrapped together with ß-galactosidase can ferment lactose.

Table 35. Data from the fermentation trials. All data are given in ml CO_2 set free at 30° C

Time	Entrapped preparation		Co-entrapped preparation		Native yeast	
	on glucose	on lactose	on glucose	on lactose	on glucose	on lactose
min	Var. 1 G	Var. 1 L	Var. 2 G	Var. 2 L	Var. 3 G	Var. 3 L
0	0 ml	0 ml	0 ml	0 ml	0 ml	0 ml
15	5 ml	0 ml	5 ml	5 ml	15 ml	0 ml
30	45 ml	0 ml	35 ml	30 ml	70 ml	0 ml
45	80 ml	0 ml	80 ml	75 ml	115 ml	0 ml
60	120 ml	0 ml	115 ml	105 ml	160 ml	0 ml
75	160 ml	0 ml	150 ml	125 ml	210 ml	0 ml
90	205 ml	0 ml	180 ml	145 ml	255 ml	0 ml
105	230 ml	0 ml	230 ml	160 ml	290 ml	0 ml
120	280 ml	0 ml	260 ml	170 ml	350 ml	0 ml
135	315 ml	0 ml	305 ml	175 ml	385 ml	0 ml
150	345 ml	0 ml	320 ml	185 ml	425 ml	0 ml
165	370 ml	0 ml	350 ml	195 ml	465 ml	0 ml
180	400 ml	0 ml	375 ml	200 ml	495 ml	0 ml
195	420 ml	0 ml	400 ml	205 ml	525 ml	0 ml
210	430 ml	0 ml	415 ml	215 ml	525 ml	0 ml
125	440 ml	0 ml	430 ml	220 ml	530 ml	0 ml
240	450 ml	0 ml	440 ml	225 ml	530 ml	0 ml

Further Evaluation and Discussion

The results shown in Table 35 become much clearer if a graphical evaluation is made (as shown in Fig. 105 for each test in which ethanol was formed). This double representation, not normally permissible in scientific publications, serves here the purpose of introducing the student to the usual steps involved in the evaluation of experimental results.

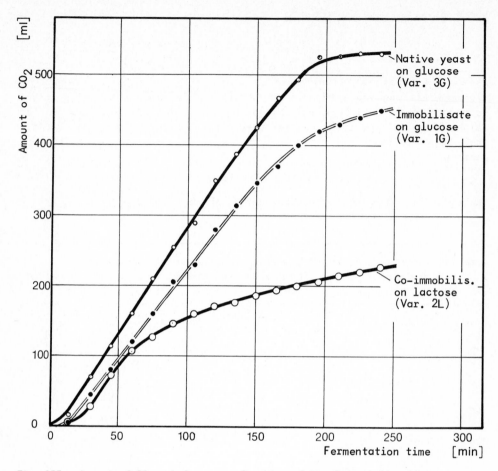

Fig. 105. Amount of CO_2 set free as a function of the fermentation time

A comparison of the fermentation curves for native (variant 3G) and entrapped cells (variant 1G) reveals that fermentation activity is slightly
reduced after entrapment. A number of factors may be responsible for this.
The alginate matrix gives rise to diffusion barriers, but in the present
experiment, even with the bead diameter of 4-5 mm, these play an insig-
nificant role because the substrate concentration outside the beads is
very high (80 g/l) and the K_S value for glucose very low (only a few
mg/l). Even if the glucose concentration drops steeply toward the center
of the bead it is still considerably above the K_S value, thus permitting
high fermentation rates.

Theoretically, partial cell damage due to the entrapment procedure
could be responsible for the reduction in fermentation activity, but this
definitely does not apply in the case of the alginate entrapment of yeast.
It is more likely that the losses are due to the nonquantitative transfer
of the entrapment mixture to the alginate beads, since, inevitably, some

of the viscous sodium alginate-yeast mixture is left in the beakers, dropping device, etc.

A conspicuous result is that lactose fermentation with the co-entrapped preparations (variant 2L) begins at quite a rapid rate but then increasingly slows down. The main reason for this is probably the competitive inhibition of the ß-galactosidase by galactose set free from lactose. As long as glucose is present in the medium, galactose is not fermented by *Saccharomyces cerevisiae*. The breakdown of galactose is, so to say, catabolically inhibited and proceeds only after a phase of adaptation during which glucose is completely fermented. Theoretically, the amount of ethanol formed can be deduced from the quantity of CO_2 measured, since it can be assumed that, under the conditions chosen, alcoholic fermentation is the only metabolic activity of yeast that sets free CO_2. The calculation can be made with the help of the well-known fermentation equation

$$C_6H_{12}O_6 \longrightarrow 2\ C_2H_5OH + 2\ CO_2$$
$$180\ g \qquad\qquad 2 \cdot 46\ g \qquad 2 \cdot 22.4\ 1$$

As an example, the step by step calculation of the specific and volumetric productivities of native and alginate-entrapped yeast on glucose substrate are shown in Table 36, starting with values taken from the steepest portion of the relevant CO_2 production curves (variants 1G and 3G) in Fig. 105 (roughly 30-150 min of fermentation).

Table 36. Calculation of the specific productivity of the native and entrapped yeast on glucose substrate

			Native yeast	Entrapped yeast
1	CO_2 set free in the trial	ml/h	180	154
2	Equivalent amount at $0°$ C	ml/h	162	138
3	Equivalent amount of ethanol	g/h	0.33	0.28
4	Specific productivity	g/g h	0.22	0.19
5	Volumetric productivity	g/l h	4.7	4.0

The figures in Table 36 are arrived at as follows:

(1) is the slope of the steepest portion of the relevant fermentation curve.

(2) is the normal gas volume at $0°$ C (273 K), which can be calculated from the value obtained at $30°$ C (303 K), by multiplying by 273 and dividing by 303.

(3) is obtained from (2) and from the relationship between the quantities of CO_2 and ethanol formed, as given by the fermentation equation.

(4) is calculated from (3), by dividing by 1.5 the dry substance (in g) of the yeast present in the fermentation experiment, where it is assumed that the usual dry substance content of 30 % applies to the 5 g compressed yeast employed.

(5) is obtained by dividing (3) by 0.07, which is the total volume, in liters, of the fermentation experiment, consisting of 50 ml substrate and 20 ml alginate beads.

On account of a number of inaccuracies the values in Table 36 are only approximate, i.e., the quantity of dry yeast substance was only roughly estimated, and, strictly speaking, the gas volume should be subjected to a pressure correction by converting from actual to normal pressures. However, the deviation resulting from this omission is unimportant compared with the error caused by CO_2 dissolving in the substrate and in the water layer sealing the gas cylinder.

Despite these and other inaccuracies, the results still allow us to draw useful comparisons. The specific productivity of the yeast in its native and entrapped form is only moderate to low as compared with good distiller's yeast, which can reach values of about 1 g/g·h. The values for the volumetric productivity in Table 36, on the other hand, are considerably higher than in the classical batchwise methods for yeast fermentation. Cell densities of this order of magnitude cannot usually be maintained using native cells and can be regarded as a characteristic advantage of immobilization.

Exercise 6   Construction and Use of a Biochemical Electrode for Assaying Glucose

E 6.1 Introduction

Using glucose oxidase and catalase, and a commercially available glass electrode (pH electrode), a biochemical electrode for measuring glucose will be constructed. From a series of measurements in which the parameters are glucose concentration and measuring time, it is intended to obtain values indicating the range of sensitivity and the response time of the enzyme electrode.

The functioning of biochemical electrodes in general and of the enzyme electrode in particular have already been discussed in Sect. 6.3. For the present exercise we have deliberately chosen a simple form of electrode, which nevertheless adequately illustrates the potentialities of, and problems connected with, the use of biosensors.

A mixture of the enzymes glucose oxidase (GOD) and catalase (CAT) is less often used in analytical procedures than catalase-free glucose oxidase. The following reactions are catalyzed:

$$\text{Glucose} + O_2 + H_2O \xrightarrow{\text{GOD}} \text{Gluconic acid} + H_2O_2$$

$$H_2O_2 \xrightarrow{\text{CAT}} H_2O + 1/2\ O_2$$

$$\text{Glucose} + 1/2\ O_2 \xrightarrow{\text{GOD+CAT}} \text{Gluconic acid}$$

As a rule, the hydrogen peroxide that accumulates in the absence of catalase is determined by means of an indicator reaction, e.g., with redox dyes. For coupling with a pH electrode, however, the mixed enzyme is better suited than catalase-free enzyme in order to prevent, or at least reduce, the inhibition of enzyme activity by peroxide.

In the foodstuff industry glucose oxidase is employed for removing oxygen from soft drinks and canned foods, as well as glucose from egg products. In such processes, catalase is necessary for preventing the accumulation of hydrogen peroxide.

E 6.2 Experimental Procedure

Construction of the Enzyme Electrode

Of a catalase-rich solution of glucose oxidase (Glucox[R] RF, John & E. Sturge Ltd., Selby, GB) 10 ml are pipetted into a Petri dish. A membrane filter is soaked in this solution for about 1 min and the excess allowed to drip off before transfer of the filter to a 300 ml Erlenmeyer flask containing a mixture of 35 ml i-propanol, 20 ml distilled water and 2 ml 25 % glutaraldehyde. The flask, containing membrane filter and reaction mixture, is now shaken for 60 min at 25° C. The membrane is then thoroughly rinsed with distilled water and kept in a storage solution in the refrigerator until required for the electrode.

The storage solution is prepared as follows: 60 g glycerol, 18 g sorbitol (Karion[R] F, E. Merck, D-6100 Darmstadt, FRG) and 0.2 ml of a disinfectant containing quaternary ammonium bases (Absonal[R], Boehringer Ingelheim KG, D-6507 Ingelheim, FRG) are made up to a volume of 100 ml with 0.05 M citrate-0.1 M phosphate buffer (pH 5.0)

Immediately before applying it to a flat-tipped electrode, the membrane is very carefully rinsed, first with distilled water and then with buffer (0.001 M sodium phosphate buffer with 0.1 M Na_2SO_4, pH 6.9). Great care must be taken not to damage the membrane and to ensure that it lies absolutely smoothly on the proton-sensitive region of the electrode (cf. Fig. 106). Finally, the membrane is secured by a rubber ring. This is best done by slipping the ring onto the open end of a test tube, as near as possible to the top. The test tube should have a smooth rim and a sufficiently large internal diameter to allow the electrode, together with its covering, to be pushed into it. After pushing the electrode into the test tube the rubber ring is slipped from the tube onto the membrane to give the arrangement shown in Fig. 106.

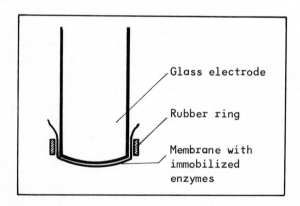

Glass electrode

Rubber ring

Membrane with immobilized enzymes

Fig. 106. Electrode covered with an enzyme-containing membrane

Measuring Procedure

In making the measurements the following conditions must be strictly observed:

> constant magnetic stirring speed of 200 min^{-1},
> constant depth of immersion of the electrode,
> constant sample volume of 100 ml,
> constant ionic strength of the test solutions,
> a pH value optimal for glucose oxidase.

A series of glucose solutions of molarities 1, 2, 3, and 5 mM are prepared in 0.001 M sodium phosphate buffer with 0.1 M Na_2SO_4 (pH 6.9). Before measurements are made, the glucose solution is thoroughly shaken in air to supply the solution with sufficient oxygen. A timer is started when the enzyme eletrode is immersed in the magnetically stirred test solution, and the pH value is then noted at suitable intervals of 1 to 5 min for 25 min. Before another sample is measured, the electrode is allowed to equilibrate in stirred pure buffer for at least 10 min.

E 6.3 Results and Evaluation

Measured Data

The values obtained fluctuate strongly from one enzyme electrode to another, so that calibrations and measurements are only comparable if made with one and the same electrode.

Table 37. pH values and pH changes measured

Time min	0.001 M Glucose		0.002 M Glucose		0.003 M Glucose		0.005 M Glucose	
	pH value	$\triangle$pH	pH value	$\triangle$pH	pH value	$\triangle$pH	pH value	$\triangle$pH
0	6.64	–	6.65	–	6.63	–	6.62	–
1	6.45	0.19	6.45	0.20	6.20	0.43	6.10	0.52
2	6.31	0.33	6.11	0.54	5.70	0.93	5.55	1.07
3	6.22	0.42	5.81	0.84	5.40	1.23	5.21	1.41
4	6.17	0.47	5.73	0.92	5.26	1.37	5.12	1.50
5	6.14	0.50	5.58	1.07	5.15	1.48	5.01	1.61
8	6.09	0.55	5.48	1.17	5.08	1.55	4.89	1.73
10	6.08	0.56	5.44	1.21	5.06	1.57	4.80	1.82
15	6.07	0.57	5.43	1.22	5.01	1.62	4.73	1.89

Graphical Representation

As Table 37 shows, the pH rises with increasing concentration of glucose: the graphs in Fig. 107 show an at least approximate proportionality over the range of 0 to 3 mmol/l. It is obvious that the enzyme electrode as constructed in this exercise has a very high response time: at glucose concentrations between 1 and 3 mmol/l, more or less constant values were only reached after about 10 min. This is probably due to the relatively thick membrane layer and means that either very long measuring times have to be tolerated, or the measuring time has to be standardized.

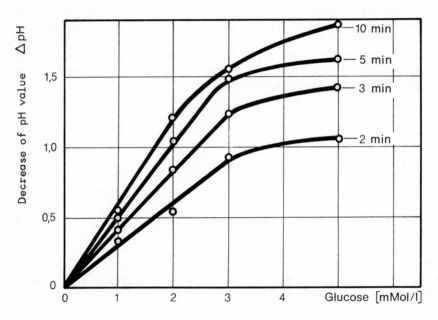

Fig. 107. pH-changes as a function of glucose concentration with duration of the measurement as parameter

Our relatively primitive example has demonstrated that it is possible to construct an enzyme electrode. The limitations that have come to light can be overcome to some extent by using a more refined measuring device and a better system. In no case can a biochemical electrode be expected to be as uncomplicated as a simple pH- or oxygen electrode.

Exercise 7 — Spinning of ß-Galactosidase into Cellulose Acetate Fibers

E 7.1 Introduction

This exercise is intended to familiarize the experimenter with the method of spinning enzymes into fibers, a special form of enzyme entrapment in a matrix (see Sect. 2.5). The example chosen is the entrapment of yeast ß-galactosidase, a procedure that is also used industrially to a a small extent.

Immobilization by this method involves entrapping fine droplets of an aqueous solution of the enzyme in the microcavities of semipermeable fibers, whereas in most other matrix-entrapment methods the enzymes are dispersed in the matrix in molecular form. The fiber material must be sufficiently dense to retain the enzymes, but at the same time porous enough to permit substrate and products to diffuse freely in both directions. Fiber-entrapped enzymes are unsuitable for the breakdown of high molecular substrates such as proteins or starches.

The starting point for all methods so far used for entrapping enzymes in fibers is the solution of a water-insoluble organic polymer in a solvent that is immiscible with water. This solution is then emulsified with an aqueous solution of the enzyme or with a cell suspension. The emulsion is then extruded through a fine nozzle into a coagulationg bath in which the polymer precipitates and hardens as a filament.

A number of enzymes have already been successfully immobilized by entrapment in fibers, but of these only ß-galactosidase from yeast has so far achieved any degree of importance in industrial processes (see Sect. 5.7). The breakdown of lactose to the monosaccharides glucose and galactose renders milk and milk products suitable for consumption by those suffering from a deficiency of ß-galactosidase and thus unable to tolerate large amounts of lactose. Up to the present time this has been the main field of application for the fiber-entrapped ß-galactosidase from yeast.

In Exercise 4 ß-galactosidase of mold origin was immobilized by cross-linking (cf. p. 154 ff). The pH optimum of the mold enzyme, between 4 and 5, corresponds to the pH range of acid whey, large quantities of which are produced in the course of cheese making, whereas the pH optimum of the yeast ß-galactosidase used in this exercise is 6 to 6.5, and thus particularly well suited for the treatment of milk, which has similar pH values.

E 7.2 Experimental Procedure

Spinning Apparatus

Figure 108 shows the apparatus for entrapping enzymes in fibers. The vessel in which emulsification takes place is a jacketed (for cooling) round bottom flask with an internal diameter of 7.5 cm. The stirrer shaft runs on two stainless steel ball bearings. All washers and stoppers are made of teflon. The stirrer is driven, via a rubber ring, by a motor that can be run at speeds varying from 0 to 2000 rpm. The coagulating bath can be cooled. The stainless steel drum used for pulling out and winding the extruded fibers has a diameter of 13 cm and a width of 5.3 cm, and is connected via a rubber ring to a motor whose speed can be varied from 0 to 200 rpm. To provide a better hold for the fibers, the drum is coated with a teflon spray before use.

Figure 108 also shows the spinneret construction. A stainless-steel, screw-on housing contains two teflon rings, the stainless mesh disk and the spinneret itself, which is a thin stainless steel disk with a centrally situated hole, 100 um in diameter.

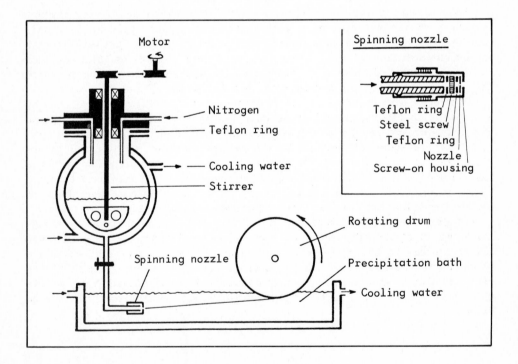

Fig. 108. Device for spinning enzymes into fibers

Spinning Procedure

In the open round flask 4 g cellulose triacetate and 50 ml methylene chloride are placed. The teflon washer is put into place and the stirrer is screwed onto the flask. Until the cellulose triacetate is totally dissolved the mixture is stirred slowly. The flask is then cooled to 0° C and, using a syringe, 8 ml of a commercial yeast ß-galactosidase (Hydro-lact[R] L50, John & E. Sturge Ltd., Selby, GB) are introduced through a special inlet provided for this purpose. After all openings have been sealed the contents of the flask are stirred at 2000 rpm for 5 min during which time a fine, milky emulsion is formed. The flask is then allowed to stand for 20 min to allow air bubbles to escape from the emulsion.

For the actual spinning procedure, nitrogen is led into the flask at a pressure of 3000 Pa (0.3 bar) and the reaction mixture is forced via the spinneret into the coagulating bath of toluene. As the polymer fiber is extruded from the spinneret it is caught with a pincette and attached to the drum, which is rotating at a speed of about 35 rpm. The nitrogen pressure and the speed of the drum are so adjusted that the fiber is extruded evenly.

When all the emulsion has been spun the fiber is removed from the drum and dried in air. The apparatus is dismantled and thoroughly washed in acetone.

Use of the Fibers

The enzyme-containing fiber is cut into lengths of approximately 2 cm and tightly packed into a jacketed glass column reactor (internal diameter 1 cm, height 10 cm), as shown in Fig. 109.

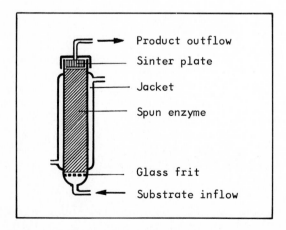

Fig. 109. Column reactor with spun ß-galactosidase

The temperature is adjusted to 40° C and a 4 % solution of lactose in 0.1 M phosphate buffer (pH 6.2) is pumped through the reactor from the bottom. The pumping rate (flow rate) is maintained at any one value (e.g., 20, 50, 100 or 200 ml/h) for 1 h. Shortly before altering the rate of flow a sample of the effluent is taken and tested for glucose by the hexokinase method (glucose test combination No. 716.251, Boehringer Mannheim GmbH, D-6800 Mannheim, FRG).

E 7.3 Results and Evaluation

The decrease in lactose hydrolysis that can be expected to result from increased flow rates is clearly shown by the results in Table 38. A maximum of 2 % glucose in the effluent can be expected from the 4 % glucose solution used in this exercise, if hydrolysis is complete. At a flow rate of 10 ml/h, 1.8 % glucose was obtained, i.e., 90 % lactose hydrolysis, which in the industrial processing of milk is considered sufficient for preventing the appearance of lactose intolerance.

The seemingly very low flow rates have to be related to the reactor volume of 7.85 ml. The resulting specific flow rates shown in the middle column of Table 38 are well within the usual range of values for immobilized biocatalysts.

Table 38. Results from measurements with spun β-galactosidase

Flow rate	Dilution rate	Glucose in the effluent
10 ml/h	1.27 h^{-1}	1.80 g/100 ml
20 ml/h	2.55 h^{-1}	1.62 g/100 ml
50 ml/h	6.37 h^{-1}	0.83 g/100 ml
100 ml/h	12.74 h^{-1}	0.41 g/100 ml

Further conclusions cannot be drawn from the present investigation since only few measurements were made, and a synthetic solution was employed. However, there is no reason why the experimenter should not extend his or her observations by making more measurements and using milk as a substrate.

Exercise 8   Immobilization of L-Asparaginase in Nylon Microcapsules

E 8.1   Introduction

The object of this exercise is to introduce the student to the interesting technique of microencapsulation of enzymes. The example chosen, i.e., the microencapsulation of L-asparaginase by boundary-layer polymerization, will show that satisfactory yields of activity can be obtained by this method. At the same time it will become clear that the relative difficulty of the encapsulation procedure and the lability of the microcapsules are considerable obstacles to their widespread, large-scale use.

Since the mid 1950's microencapsulated dyes have been used to produce "self-inking" papers. In this technique microcapsules containing a dye are incorporated into the paper. When pressure is applied, either to the carbonless copy paper itself or another paper covering it, the capsules are disrupted and the paper becomes colored at the site of pressure.

It is more likely that microencapsulated enzymes will find use in medical and analytical sectors than in industry. In the artificial kidney, for example, microencapsulated urease is employed for the breakdown of urea to carbon dioxide and ammonia (cf. Sect. 7.3). Other studies in progress are aimed at the use of microencapsulated asparaginase in treating certain forms of cancer, such as lymphosarcoma (cf. Sect. 7.1).

As an example of an enzyme with potentialities in the field of medicine, L-asparaginase will be microencapsulated in this exercise. This enzyme catalyzes the hydrolytic breakdown of L-asparagine to L-aspartic acid and ammonia. Theoretically, therefore, it should be able to eliminate the L-asparagine essential for a lymphosarcoma, thus depriving the sarcoma of the nutrient necessary for its development.

Various methods of microencapsulation have already been described in Sect. 2.6. In the present exercise we shall use boundary-layer polymerization, a very popular technique. In this method, the capsules form at the phase boundary, following emulsification of an aqueous solution of the enzyme in an organic solvent that is not soluble in water. One monomer dissolved in the aqueous phase and a second monomer dissolved in the organic solvent phase combine at their site of contact to form the capsule polymer.

E 8.2 Experimental Procedure

Microencapsulation

The L-asparaginase solution used in the experiment (product No. 102.903, Boehringer Mannheim GmbH, D-6800 Mannheim, FRG) contains 5 mg L-asparaginase from *E. coli* in 1 ml glycerol. The original solution (1 ml) is diluted to a total of 5 ml with 4 ml ice-cold distilled water. The diluted asparaginase solution (1 mg asparaginase per ml) is used in the following experiments.

Solutions 1, 2 and 3 are prepared separately in an ice bath, where they are kept in readiness for the encapsulation procedure.

Solution 1. In 2.5 ml 0.45 M borate buffer (pH 8.4) 1 ml diluted L-asparaginase solution (1 mg/ml), 1 mg casein according to Hammarsten, 7 mg aspartic acid and 93 mg hexamethylene diamine (= 1,6-diaminohexane) are dissolved.

Solution 2. 16 ml cyclohexane, 4 ml chloroform, and 2 drops of Span 85 are mixed.

Solution 3. 0.1 ml sebacoyl chloride (= sebacic acid dichloride), 12 ml cyclohexane and 3 ml chloroform are mixed.

Solutions 1 and 2 are combined in the ice bath and stirred magnetically for 3 to 4 min at a speed sufficient to produce an emulsion of droplets approximately 1 to 2 μm in diameter (checked under the microscope). Under continuous stirring, solution 3 is now slowly added (over a period of 8 to 10 min) by pipetting along the edge of the vessel, after which the emulsion is stirred for a further 2 min.

The microcapsules thus produced are separated on a glass frit, which should at no point be allowed to be sucked dry. The microcapsules are then rinsed with ethanol, and very thoroughly washed with distilled water. They are finally suspended in 10 ml 0.02 M phosphate buffer (pH 6.5). A drop of this suspension is inspected under the microscope in order to ascertain the size and shape of the capsules. The rest of the suspension is kept in the icebox until required for activity determinations.

Determination of the L-Asparaginase Activity

The activity determination is carried out according to the scheme shown in Table 39, which is largely taken from Bergmeyer et al. (1974). The activities of the encapsulated enzyme preparation and of the native enzyme are determined in this way. For testing the encapsulated enzyme 0.01 ml of the suspension from the icebox is used (corresponding to 10 μg enzyme protein). For the soluble, native enzyme, a 1 to 100 dilution of the 1 mg/ml solution (see p. 175) is made, and of this 0.1 ml (corresponding to 1 ug enzyme) is used for each test.

Table 39. Scheme for the determination of L-asparaginase activity

	Reference standard		Enzyme preparation	
	blank	standard	blank	prepar.
Tris buffer (0.2 M; pH 8.6)	1.00 ml	1.00 ml	1.00 ml	1.00 ml
L-Asparagine (25.5 mg/ml)	-	-	0.10 ml	0.10 ml
Distilled water	1.00 ml	-	0.90 ml	0.80 ml
$(NH_4)_2SO_4$-Standard (161 μg/ml)	-	1.00 ml	-	-
Soluble or suspended preparation	-	-	-	0.10 ml

Mixing and incubation 30 min at 37° C

Trichloracetic acid (1.5 M)	0.10 ml	0.10 ml	0.10 ml	0.10 ml
Distilled water	-	-	-	0.10 ml
Soluble or suspended preparation	0.10 ml	0.10 ml	0.10 ml	-

Mixing and analysis of the
supernatant in the following test

Distilled water	4.25 ml	4.25 ml	4.25 ml	4.25 ml
Supernatant	0.25 ml	0.25 ml	0.25 ml	0.25 ml
Nesslers Reagent	0.50 ml	0.50 ml	0.50 ml	0.50 ml

Mixing and 60 s   later measurement of extinction at 436 nm (1 cm layer depth)

E 8.3 Results and Evaluation

Table 40 shows the values obtained and the enzyme activities calculated
from them. The calculation is described in detail below for those less
familiar with the procedure.

Table 40. Enzyme activity of the native and encapsulated L-asparaginase

	Native L-asparaginase	Encapsulated L-asparaginase
Extinction of the reference standard	0.144	0.144
Extinction of the enzyme	0.153	0.223
Amount of protein in the test mg	0.001	0.010
Activity u/mg	86.3	12.6
Yield of the immobilization %	-	14.6

The value to be calculated is the enzyme activity in units per weight (u/mg), one unit (u) being the activity that results in the release of 1 μmol NH_4 per minute. The standard employed was $(NH_4)_2SO_4$ (molecular weight 132.14). The 161 μg used are equivalent to 1.2184 μmol $(NH_4)_2SO_4$ or 2.4368 μmol NH_4^+. The quantity of NH_4^+ released in μmol per 30 min can be found by substituting the extinction values measured for the preparation and the standard in the following equation:

$$c = \frac{E_{prep.}}{E_{stand.}} \; 2.4368 \quad \left[\mu mol \; NH_4^+/30 \; min \right]$$

The amount of NH_4^+ released, in μmol per min, is obtained by dividing this value by 30. Finally, the result has to be divided by the quantity of enzyme originally used, in mg, to give the amount of NH_4^+ released per mg original enzyme. For the native preparation this was 1 μg (0.001 mg), and hence its activity is given by

$$A = \frac{E_{prep.}}{E_{stand.}} \cdot \frac{2.4368}{30 \cdot 0.001} \quad \left[u/mg \right]$$

Of the microencapsulated enzyme 10 μg were taken (0.01 mg) and its activity is thus given by

$$A = \frac{E_{prep.}}{E_{stand.}} \cdot \frac{2.4368}{30 \cdot 0.010} \quad \left[u/mg \right]$$

The 15 % yield of activity achieved by encapsulation is satisfactory, and within the range of values reported in the literature. However, considerably greater precautions were taken to protect the enzyme in this case (short exposure to the monomers, icebath, etc.) than in the other methods of immobilization practiced.

Exercise 9   Degradation of Cellulose in a Membrane Reactor

E 9.1   Introduction

As described in Sect. 5.3, enzymes in membrane reactors are already being used industrially for the production of amino acids. The basic techniques involved in employing membrane reactors will be demonstrated here in the breakdown of a polysaccharide, this being a simpler process than the production of amino acids. The way in which a membrane reactor works is very easily demonstrated using the enzymatic breakdown of cellulose, a homopolymer of glucose, as an example. The low-molecular products can cross the membrane and are then estimated as reducing sugars. Enzymes and high-molecular, or even insoluble, polysaccharides are held back by the membrane.

In addition to acquainting the student with the membrane reactor, the present exercise gives an idea of some of the basic problems connected with enzymatic degradation of cellulose, which is one of the most thoroughly investigated fields of enzyme technology throughout the world. At the back of such intensive research is the hope of discovering better ways of utilizing the vast quantities of cellulose that are continuously provided by nature in the form of wood and straw. So far, no completely satisfactory solution, from the economic point of view, has been found.

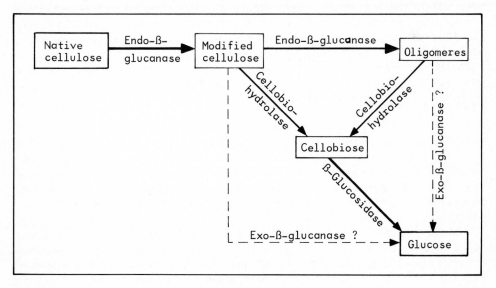

Fig. 110. Enzymatic breakdown of cellulose

In contrast to the carbohydrate storage substance starch, the ß-glucan cellulose is designed by nature to fulfil the role of a durable, structural polysaccharide: it is correspondingly difficult to break down. Even if purified cellulose is used its enzymatic breakdown remains a difficult and complex procedure. Figure 110 shows the most important steps in cellulose degradation, together with the enzymes involved: it is clear that only by using a combination of enzymes can satisfactory hydrolysis be expected. An important aspect in this context is the synergistic effect of the different cellulolytically active components of the enzyme preparation on one another. The technical preparation of cellulase used in our experiment contains endo- as well as exo-ß-1,4-glucanases.

The procedure becomes considerably more complicated than that shown in Fig. 110 if typical cellulose-containing raw material such as wood or straw are used; their cellulose is interspersed with large amounts of lignin and hemicellulose, making necessary additional measures for their breakdown. The experiments in the form described here should therefore be regarded as models upon which more elaborate procedures can be based.

E 9.2 Experimental Procedure

Substrate

10 g cellulose (see below) are stirred until smooth and homogeneous with a small volume of 0.1 M citrate-phosphate buffer, pH 4, before making up to a volume of 1000 ml with the same buffer, to which 0.2 g sodium azide have been added. The experiments are performed with microcrystalline cellulose (Sigmacell^R, product No. S 3755, Sigma, D-8024 Deisenhofen, FRG) and with a cellulose derivative, carboxymethyl cellulose (CM-cellulose Na salt, product No. 9004-32-4, Fluka, D-7910 Neu-Ulm, FRG).

Experimental Set-up

In the arrangement shown in Fig. 111 an ultrafiltration cell with a volume of 65 ml is used; the ultrafilter membranes (Diaflow^R PM 10, Amicon GmbH, D-5810 Witten, FRG) are 43 mm in diameter. To prevent dissolved substrate from settling, the cellulose is continuously stirred. The membrane chamber is stirred at 800 rpm. Substrate additions are made by means of a peristaltic pump (type No. 101U/R, Watson Marlow, Falmouth, GB)

The experiments are carried out at 30° C in an incubation chamber. 1.5 g cellulase (Cellulase CP, John & E. Sturge Ltd., Selby, GB) dissolved or suspended in approximately 50 ml substrate are placed in the membrane reactor (outflow closed) and the volume made up to the full 65 ml. After stirring for 2 h the outflow is opened and substrate is pumped in at a flow rate of about 32.5 ml/h, the ultrafiltrate leaving at the same speed. The exact flow rate must be measured. Samples (2-3 ml) of effluent are collected at intervals of 1 h or more, inactivated by heating, and stored until analyzed for reducing sugar. The experiment is to be performed with both substrates mentioned above, and also at additional flow rates, higher and lower than 32.5 ml/h.

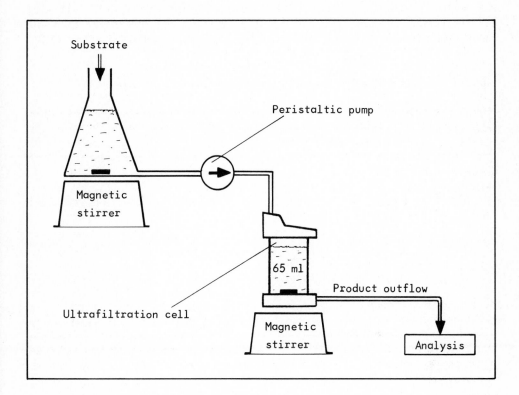

Fig. 111. Experimental equipment

Determination of Reducing Sugar

Reducing sugar is determined according to the procedure already described in Sect. E 1.2, using dinitrosalicylic acid. A calibration curve for glucose is plotted. The results are then also expressed as g glucose per liter, although the reducing sugar is present in the form of glucose oligomers.

E 9.3 Results and Evaluation

The results obtained are assembled in Table 41. It proved impossible to raise the flow rate above 75 ml/h since the pressure in the reactor then became too high. The reducing sugar is determined after roughly five times the reactor volume of substrate have flowed through the reactor following any alteration in flow rate. For example, given a reactor volume of 65 ml and a flow rate of 32.5 ml/h the reducing sugar is determined after 10 h.

Table 41. Trials with microcrystalline cellulose and CM-cellulose

Cellulose type	Flow rate ml/h	Dilution rate h^{-1}	Glucose set free g/l	Productivity g/l·h
Microcrystalline cellulose	16	0.246	1.12	0.276
	28	0.431	0.65	0.280
	45	0.692	0.72	0.498
	62	0.954	0.38	0.363
CM-cellulose	15	0.231	1.85	0.427
	22	0.338	1.50	0.508
	30	0.462	1.60	0.738
	41	0.631	1.48	0.934
	44	0.677	1.52	1.029
	64	0.985	1.27	1.251

Further Evaluation and Discussion

From the graphical representation (Fig. 112) of the data taken from Table 41 it becomes clear that the amount of substrate broken down is dependent on the flow rate. As was to be expected, the cellulose derivative CM cellulose is more easily broken down than the microcrystalline cellulose. This can be seen from Fig. 112, even if the values show considerable scatter. It is also not surprising that an increase in flow rate, with the resulting decreased residence time, results in a lower degree of breakdown. For limited periods of time anyway a continuous breakdown of pure cellulose can be achieved, but in longer experiments, and using impure technical cellulose, problems arising from membrane blockage are likely.

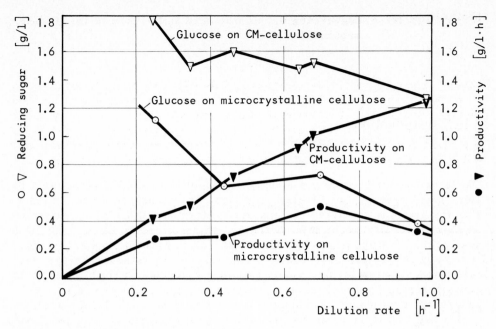

Fig. 112. Reducing sugar and productivity as a function of dilution rate

As already seen in Sect. 4.1, the volumetric productivity for one-step, continuous systems is obtained by multiplying the concentration of the product by the dilution rate. In the present case, as can be seen from Table 41, this calculation gives values of up to $P_v = 1.25$ g/l·h. This means that per m^3 reactor volume about 1.25 kg reducing sugar can be produced per hour, a value that resembles in order of magnitude the productivity values obtained by the classical batchwise fermentation procedure. Since no attempt at optimization was made here, and the experiments were performed at 30° C, the values can certainly be improved upon. Nevertheless, cellulose degradation on a large scale by means of this method would only be economically feasible if a very cheap source of microcrystalline cellulose or cellulose derivatives were to become available.

Exercise 10 Conversion of Fumaric Acid to Malic Acid in a Liquid-membrane Emulsion

E 10.1 Introduction

The technique for producing and applying liquid-membrane emulsions will be illustrated here in the fumarase reaction. As already described in Sect. 5.8, the reaction catalyzed by fumarase leads to the hydrolytic conversion of fumaric acid to malic acid. Normally, this process is carried out with dead cells; the use of a liquid membrane as an alternative will almost certainly not attain any economic importance. It is nevertheless well suited for illustrating the liquid membrane technique as such, since it is straightforward to perform and its progress can be followed by simple analytical procedures. Although multienzyme systems in liquid-membrane emulsions, including retention of cofactors (Makryaleas et al. 1985), are known from the literature, the technique as a whole is still at a preliminary stage.

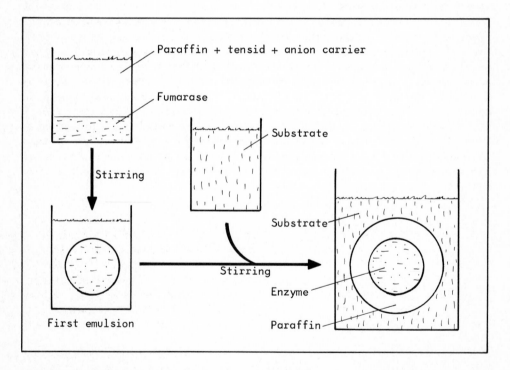

Fig. 113. Formation of liquid membrane emulsions

Basically, the enzyme-emulsion technique offers another method in addition to the liposome technique (see Sect. 2.6) for confining enzymes in a limited space by a liquid membrane. In contrast to liposomes, the liquid membrane in the emulsion technique does not consist of a regularly arranged double layer of amphilic molecules, but of a thick, hydrophobic liquid shell. The very recently developed methods for producing liquid membrane emulsions are based on the normally good solubility of enzymes in an aqueous milieu and their insolubility in hydrophobic media such as paraffin, kerosene, or chloroform.

Figure 113 shows how a liquid-membrane emulsion can be obtained. The enzyme-containing aqueous phase is first of all emulsified in the water-immiscible phase by vigorous stirring. The addition of surfactants makes it possible to achieve an emulsion of extremely fine droplets; this is one of the essential conditions for their stability during the subsequent addition of another aqueous phase containing substrate. The enzyme-containing droplets are separated from the surrounding aqueous substrate solution by a hydrophobic layer that acts as a liquid membrane. The apparatus and the reactions involved are shown schematically in Fig. 114.

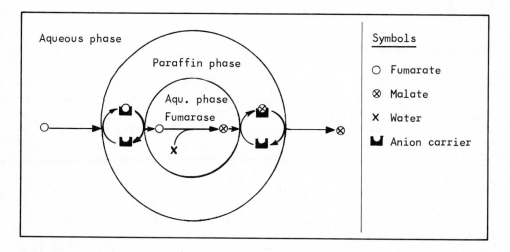

Fig. 114. Scheme of mass transfer and reactions in the enzyme-emulsion system

Anions can be transported through the paraffin layer if it contains a liquid anion exchanger. In the example described here the anion exchanger takes up fumarate ions at the outer boundary surface and releases them at the inner surface (see Fig. 114). Conversely, it can transport malate ions from the inside to the outside. These transport processes obey statistical laws. Since there are far more fumarate ions outside than inside the liquid membrane, fumarate uptake predominates at the outer boundary layer, whereas on the inside the anion carrier releases more fumarate than it takes up. Transport can only proceed in the presence of a concentration gradient.

E 10.2 Experimental Procedure

The following dilutions of fumarase (Boehringer Mannheim GmbH, D-6800 Mannheim, FRG) are used:

0.05 ml fumarase (0.1 mg) + 1.95 ml 0.2 M phosphate buffer pH 7.5,
0.1 ml fumarase (0.2 mg) + 1.9 ml 0.2 M phosphate buffer pH 7.5.

The membrane phase is produced by mixing 5 ml liquid paraffin, 0.25 ml tensid (Span 85) and one drop of an anion carrier (Adogen[R] 464, Serva, D-6900 Heidelberg, FRG). 2 ml of this membrane phase together with 2 ml diluted fumarase (see above) are homogenized for 90 s in a glass centrifuge tube using an Ultra-Turrax. The emulsion thus produced is added to 10 ml 0.1 M fumarate solution under intensive magnetic stirring. During the subsequent incubation at 25° C the mixture is stirred gently. Samples are withdrawn at hourly intervals over a period of 4 h. The sample filtrate, after appropriate dilution, is tested for L-malic acid (Test combination No. 139068, Boehringer Mannheim GmbH, D-6800 Mannheim, FRG).

Experimental Variants

In addition to the normal procedure described above, a variant is performed which differs from the normal version only in the omission of the anion carrier Adogen[R] 464.

In another experiment, the normal procedure is followed in every respect, but a sample is withdrawn following the intensive mixing of the enzyme with the fumaric acid substrate. The filtrate of this sample is incubated separately and tested for L-malic acid. In this way it is possible to determine the degree to which enzyme released by the stirring process is responsible for the formation of L-malic acid.

Experiments in which native enzyme replaces the encapsulated enzyme are aimed at giving an idea of the extent to which encapsulation reduces the speed of reaction by limiting transport. Since it can be expected that malic acid will be produced much more rapidly in the experiment with native enzyme, only 0.05 ml native fumarase should be used.

E 10.3 Results and Evaluation

A glance at the data obtained by students, and assembled in Table 42, is sufficient to reveal that conversion of fumaric acid to malic acid has taken place in the liquid membrane system. The speed of the reaction is considerably raised by the presence of Adogen^R 464, as an anion carrier, in the lipohilic membrane.

Table 42. Formation of malic acid in the different systems

Enzyme used form	amount	Anion carrier	Time	Dilution factor	Extinction values E1	E2	Malic acid
encap-sulated	0.05 ml	yes	1 h	100	0.501	0.587	3.91 g/l
		yes	2 h	100	0.504	0.629	5.66 g/l
		yes	3 h	100	0.503	0.641	6.23 g/l
		yes	4 h	100	0.501	0.656	7.00 g/l
encap-sulated	0.10 ml	yes	1 h	100	0.505	0.639	6.00 g/l
		yes	2 h	100	0.508	0.653	6.66 g/l
		yes	3 h	100	0.502	0.662	7.23 g/l
		yes	4 h	100	0.507	0.718	9.64 g/l
encap-sulated	0.10 ml	no	1 h	100	0.501	0.540	1.51 g/l
		no	2 h	100	0.500	0.566	2.79 g/l
		no	3 h	100	0.509	0.582	3.12 g/l
		no	4 h	100	0.508	0.593	3.69 g/l
native	0.05 ml	yes	1 h	1000	0.500	0.532	11.81 g/l
		yes	2 h	1000	0.507	0.542	13.23 g/l
		yes	3 h	1000	0.508	0.545	14.65 g/l
		yes	4 h	1000	0.507	0.547	15.59 g/l
filtrate	0.10 ml	yes	1 h	100	0.501	0.503	0.10 g/l
		yes	2 h	100	0.516	0.520	0.19 g/l
		yes	3 h	100	0.504	0.513	0.43 g/l
		yes	4 h	100	0.507	0.526	0.85 g/l

The results shown in Table 42 are easier to interpret if plotted as shown in Fig. 115. It becomes obvious that in each case the reaction speed decreases with increasing incubation time. To what extent this is due to increasing enzyme inactivation, product inhibition or, in the case of the encapsulated enzyme, to impeded transport, cannot be decided on the basis of the present experiments. However, from other studies using the same system it can be concluded that inactivation plays no significant role under our experimental conditions.

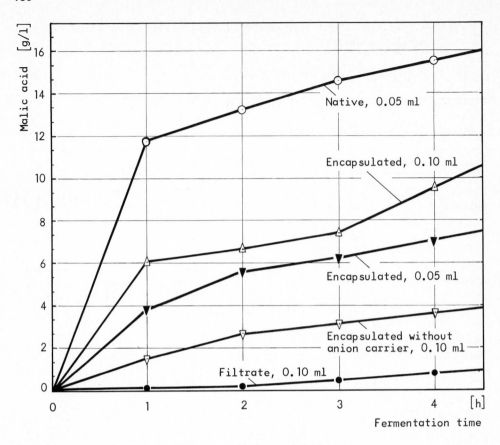

Fig. 115. Malic acid formed by the different systems used

The results also clearly illustrate several problems connected with liquid membrane systems. The lipophilic membranes are highly sensitive and easily destroyed: the occurrence of malic acid in the external phase indicates that immediately following mixing of the encapsulated enzyme with the substrate an appreciable amount of the enzyme has already escaped across the membrane. A very strong reduction in the rate of formation of reaction products following membrane encapsulation of fumarase shows that this procedure is associated with a considerable limitation of transport. Although the addition of an anion carrier brings about a marked improvement in the transport of both substrate and product, it does not raise the reaction rate to anything approaching the values obtained with the free enzyme. Liquid-membrane emulsions are far too sensitive to be suitable for use in industrial processes, although they might be of valuable use in basic research in the imitation and study of cellular compartmentalization.

APPENDIX

Abbreviations and Symbols

Abbreviations

ADP	Adenosine diphosphate
Ala	Alanine
6-APA	6-Aminopenicillanic acid
Arg	Arginine
Asn	Asparagine
Asp	Aspartic acid
ATP	Adenosine triphosphate
BIS	Bis(N,N´)-Methylenebisacrylamide
CAT	Catalase
CM-	Carboxymethyl-
CoA	Coenzyme A
Cys	Cysteine
DEAE-	Diethylaminoethyl-
EC	Enzyme Commission
EDTA	Ethylene diamine tetraacetic acid
FAD	Flavin adenine dinucleotide
Gln	Glutamine
Glu	Glutamic acid
Gly	Glycine or glycocoll
GOD	Glucose oxidase
His	Histidine
IEP	Isoelectric point
Ileu	Isoleucine
IUB	International Union of Biochemistry
Leu	Leucine
Lys	Lysine
Met	Methionine
MW	Molecular weight
NAD	(NAD^+) Nicotinamide adenine dinucleotide
$NADH_2$	($NADH + H^+$) reduced nicotinamide adenine dinucleotide
NADP	($NADP^+$) Nicotinamide adenine dinucleotide phosphate
NC	Nomenclature Commission
ONPG	o-Nitrophenyl-ß-D-galactopyranoside
PAL	Pyridoxal phosphate
PEG	Polyethylene glycol
Phe	Phenylalanine
Pro	Proline
Ser	Serine
Thr	Threonine

TPP	Thiamine pyrophosphate
Trp	Tryptophane
Tyr	Tyrosine
Val	Valine

Symbols

The dimensions given to the symbols in the following list are only examples. Depending on the context, other dimensions are sometimes more appropriate. The amount of a substance can, for example, be given in g, kg, mol, mmol, μmol, etc.

A	(u/g)	Activity
$\bar{A}$	(g/l)	Product concentration
A_o	(u/g)	Activity at time 0
A_t	(u/g)	Activity at time t
COD	(g/l)	Chemical oxygen demand
D	(h^{-1})	Specific flow rate, dilution rate
D_e	(cm^2/s)	Effective diffusion constant
e	(-)	Basis of the natural logarithm
E	(-)	Extinction
ε	($cm^2/\mu mol$)	Extinction coefficient
E_a	(kJ/mol)	Activation energy
E_{ao}	(kJ/mol)	Activation energy without biocatalyst
η	(-)	Effectivity
f	(1/h)	Flow rate
F	(cm^2)	Surface area
ΔG	(kJ/mol)	Free energy of a reaction
h	(m)	Height, layer depth
k	(h^{-1})	Inactivation coefficient
K_m	(mol/l)	Michaelis constant
$K_{m,s}$	(mol/l)	Apparent Michaelis constant
k_p	(Pa/m)	Coefficient of pressure buildup
K_s	(mol/l)	Monod constant
μ	(m^2/s)	Viscosity
P	(cm^3/s)	Permeability factor
ΔP	(Pa)	Pressure difference, pressure buildup
Pv	(g/l·h)	Volumetric productivity
φ	(-)	Thiele modulus
r	(cm)	Diffusion distance, radius
R	(J/mol·K)	General gas constant
S	($mmol/cm^3$)	Substrate concentration
ΔS	($mmol/cm^3$)	Difference of substrate concentration
S_{ex}	($mmol/cm^3$)	Substrate concentration outside

S_{en}	$(mmol/cm^3)$	Substrate concentration inside
Sh	(-)	Sherwood number
T	(K)	Absolute temperature
T	(mol)	Amount of enzymatic turnover
t	(h)	Time, duration
$t_{1/2}$	(h)	Half-life
t_m	(h)	Average residence time
v	(mmol/s)	Reaction velocity, conversion rate
v_d	(mmol/s)	Diffusion velocity, diffusion rate
v_o	(m/s)	Surface velocity
V	(1)	Volume
V_{max}	(mmol/s)	Maximal reaction velocity

Literature

The number of publications on immobilized biocatalysts that has appeared in recent decades is so vast that only a very small selection can be mentioned within the framework of an introductory text such as this. In addition to the references cited, some summarizing books are listed below. As a guide to further reading, some of the most recent original publications and survey articles relevant to each of the main chapters are presented. To assist the reader in deciding whether or not an article is of interest to him, all references are given with their full title.

Books

Buchholz K (ed) (1979) Characterization of immobilized biocatalysts, Dechema monographs, vol **84**. Verlag Chemie, Weinheim, 394 pp

Chibata I (ed) (1978) Immobilized enzymes research and development. Wiley, New York, 284 pp

Chibata I, Wingard L B jr (eds) (1983) Applied biochemistry and bioengineering, vol **4**, immobilized microbial cells. Academic Press, New York, 355 pp

Ghose T K, Fiechter A, Blakebrough N (eds) (1978) Advances in biochemical engineering, vol **10**, immobilized enzymes I. Springer, Berlin Heidelberg New York, 177 pp

Ghose T K, Fiechter A, Blakebrough N (eds) (1979) Advances in biochemical engineering, vol **12**, immobilized enzymes II. Springer, Berlin Heidelberg New York, 253 pp

Laskin A I (ed) (1985) Applications of isolated enzymes and immobilized cells to biotechnology. Adison-Wesley, Amsterdam, 300 pp

Mattiasson B (ed) (1983) Immobilized cells and organelles, vol **1**. CRC Press, Boca Raton, 143 pp

Mattiasson B (ed) (1983) Immobilized cells and organelles, vol **2**. CRC Press, Boca Raton, 158 pp

Mosbach K (ed) (1976) Methods in enzymology, vol **44**, immobilized enzymes. Academic Press, New York, 999 pp

Webb C, Black G M, Atkinson B (1986) Process engineering aspects of immobilized cell systems. Institution Chemical Engineers, Rugby, 240 pp

Wingard L B jr, Katchalski-Katzir E, Goldstein L (eds) (1981) Applied biochemistry and bioengineering, vol **3**, analytical applications of immobilized enzymes and cells. Academic Press, New York, 336 pp

Woodward J (ed) (1985) Immobilized cells and enzymes, a practical approach. IRL-Press, Oxford, 192 pp

Literature Cited

Ahmed F, Dunlap R B (1984) Kinetic studies of sepharose- and CH-sepharose-immobilized dihydrofolate reductase. Biotechnol Bioeng 26: 1227-1232

Aizawa W, Wada M, Kato S, Suzuki S (1980) Immobilized mitochondrial electron transport particle for NADH determination. Biotechnol Bioeng 22: 1769-1783

Angelino S A G F, Müller F, Plas H C van der (1985) Purification and immobilization of rabbit liver aldehyde oxidase. Biotechnol Bioeng 27: 447-455

Bachmann S, Gebicka L, Gasyna Z (1981) Immobilization of glucose isomerase on radiation-modified gelatine gel. Starch/Stärke 33: 63-66

Banerjee M, Chakravarty A, Majumdar S K (1984) Characteristics of yeast β-galactosidase immobilized on calcium alginate gels. Appl Microbiol Biotechnol 20: 271-274

Bar R, Gainer J L, Kirwan D J (1986) Immobilization of Acetobacter aceti on cellulose ion exchangers: adsorption isotherms. Biotechnol Bioeng 28: 1166-1171

Barbaric S, Kozulic B, Leustek I, Pavlovic B, Cesi V. Mildner P (1984) Cross-linking of glycoenzymes via their carbohydrate chains. In: 3rd Eur Congr Biotechnol, vol 1. Verlag Chemie, Weinheim, pp 307-312

Beddows C G, Guthrie J T, Abdel-Hay F I (1981) The use of graft copolymers as enzyme supports for immobilization of proteins and enzymes on a hydrolyzed nylon-co-acrylnitrile system. Biotechnol Bioeng 23: 2885-2889

Beddows C G, Gil M H, Guthrie J T (1982) The immobilization of enzymes, bovine serum albumin, and phenylpropylamine to poly(acrylic acid)-polyethylene-based copolymers. Biotechnol Bioeng 24: 1371-1387

Bergmeyer H U, Gawehn K, Graßl M (1974) L-Asparaginase. In: Bergmeyer H U (ed) Methoden der enzymatischen Analyse, 3rd ed, vol 1. Verlag Chemie, Weinheim, pp 464-465

Bettmann H, Rehm H J (1984) Degradation of phenol by polymer entrapped microorganisms. Appl Microbiol Biotechnol 20: 285-290

Bihari V, Goswami P P, Rizvi S H M, Kahn A W, Basu S K, Vora V C (1984) Studies on immobilized fungal spores of microbial transformation of steroids: 11a-hydroxylation of progesterone with immobilized spores of Aspergillus ochraceus G8 on polyacrylamide gel and other matrices. Biotechnol Bioeng 26: 1403-1408

Black G M, Webb C, Matthews T M, Atkinson B (1984) Practical reactor systems for yeast cell immobilization using biomass support particles. Biotechnol Bioeng 26: 134-141

Boudrant J, Ceheftel C (1975) Continuous hydrolysis of sucrose by invertase adsorbed in a tubular reactor. Biotechnol Bioeng 17: 827

Brouers M, Hall D O (1986) Ammonia and hydrogen production by immobilized cyanobacteria. J Biotechnol 3: 307-321

Cabral J M S, Novais J M, Cardoso J P (1984) Coupling of glucoamylase on alkylamine derivative of titanium(IV) activated controlled pore glass with tannic acid. Biotechnol Bioeng 26: 386-388

Cannon J J, Chen L-F, Flickinger M C, Tsao G T (1984) The development of an immobilized lactate oxidase system for lactic acid analysis. Biotechnol Bioeng 26: 167-173

Cantarella M, Migliaresi C, Tafuri M G, Alfani F (1984) Immobilization of yeast cells in hydroxymethacrylate gels. Appl Microbiol Biotechnol 20: 233-237

Chipley J R (1974) Effects of 2,4-dinitrophenol and N,N'-dicyclohexylcarbodiimide on cell envelope-associated enzymes of Escherichia coli and Salmonella enteritidis. Microbios 10: 115-120

Clark D S, Bailey J E (1984) Deactivation kinetics of immobilized a-chymotrypsin subpopulations. Biotechnol Bioeng 26: 1090-1097

Cocquempot M F, Thomasset B, Barbotin J N, Gellf G, Thomas D (1981) Comparative stabilization of biological photosystems by several immobilization procedures. 2. Storage and functional stability of immobilized thylakoids. Eur J Appl Microbiol Biotechnol 11: 193-198

Cohenford M A, Santaro P F, Urbanowski J C, Dain J A (1986) Effect of immobilization on the physical and kinetic properties of soluble and insoluble trypsin-albumin polymers. Biotechnol Bioeng 28: 736-740

D'Angiuro L, Cremonesi P (1982) Immobilization of glucose oxidase on sepharose by uv-initiated graft copolymerization. Biotechnol Bioeng 24: 207-216

Decleire M, Huyhn N van, Motte J C (1985) Hydrolysis of lactose solutions and wheys by whole cells of Kluyveromyces bulgaricus. Appl Microbiol Biotechnol 21: 103-107

Deo Y M, Gaucher G M (1984) Semicontinuous and continuous production of penicillin-G by Penicillium chrysogenum cells immobilized in k-carrageenan beads. Biotechnol Bioeng 26: 285-295

De Rosa M, Gambacorta A, Lama L, Nicolaus B (1981) Immobilization of thermophilic microbial cells in crude egg white. Biotechnol Lett 3: 183-188

DiLuccio R C, Kirwan D J (1984) Effect of dissolved oxygen on nitrogen fixation by A. vinelandii. II. Ionically adsorbed cells. Biotechnol Bioeng 26: 87-91

Döppner T, Hartmeier W (1984) Glucose oxidation by modified mould mycelium. Starch/Stärke 36: 283-287

Dua R D, Kumar S, Vasudevan P (1985) Caboxypeptidase A immobilization on activated styrene-malic anhydride systems. Biotechnol Bioeng 27: 675-680

Dwyer D F, Krumme M L, Boyd S A, Tiedje J M (1986) Kinetics of phenol biodegradation by an immobilized methanogenic consortium. Appl Environm Microbiol 52: 345-351

Ehrhardt H M, Rehm H J (1985) Phenol degradation by microorganisms adsorbed on activated carbon. Appl Microbiol Biotechnol 21: 32-36

Eikmeier H, Rehm H J (1984) Production of citric acid with immobilized Aspergillus niger. Appl Microbiol Biotechnol 20: 365-370

Förberg C, Häggström L (1984) Adsorbed cell systems controlled by the nutrient dosing technique. In: 3rd Eur Congr Biotechnol, vol 2. Verlag Chemie, Weinheim; pp 115-120

Fradet H, Arnaud A, Rios G, Galzy P (1985) Hydration of nitriles using a bacterial nitrile-hydratase immobilized on DEAE-cellulose. Biotechnol Bioeng 27: 1581-1585

Friend B A, Shaghani K M (1982) Characterization and evaluation of Aspergillus oryzae coupled to a regenerable support. Biotechnol Bioeng 24: 329-345

Fukushima S, Yamade K (1982) Rapid continuous alcohol fermentation of carbohydrates in a novel immobilized bioreactor. In: Dellweg H (ed) 5th Symp Techn Mikrobiol. VLSF, Berlin, pp 346-354

Gainer J L, Kirwan D J, Foster J A, Seylan E (1980) Use of adsorbed and covalently bound microbes in reactors. Biotechnol Bioeng Symp 10: 35-42

Garde V L, Thomasset B, Tanaka A, Gellf G, Thomas D (1981) Comparative stabilization of biological photosystems by several immobilization systems. 1. ATP production by immobilized bacterial chromatophores. Eur J Appl Microbiol Biotechnol 11: 133-138

Gianetto A, Berruti F, Glick B R, Kempton A G (1986) The production of ethanol from lactose in a tubular reactor by immobilized cells of Kluyveromaces fragilis. Appl Microbiol Biotechnol 24: 277-281

Giard D J, Loeb D H, Thilly W G, Wang D I C, Levine D W (1979) Human interferon production with diploid fibroblast cells grown on microcarriers. Biotechnol Bioeng 21: 433-442

Greenberg N A, Mahoney R A (1981) Immobilization of lactase (ß-galactosidase) for use in dairy processing: a review. Process Biochemistry 16 (no 2): 2-8 and 49

Haecht J L van, Bolipombo M, Rouxhet P G (1985) Immobilization of Saccharomyces cerevisiae by adhesion: treatment of the cells by Al ions. Biotechnol Bioeng 27: 217-224

Hartmeier W (1977) Immobilisierte Enzyme für die Lebensmitteltechnologie. Gordian 77: 202-210, 232-237

Hartmeier W, Tegge G (1979) Versuche zur Glucoseoxidation in Glucose-Fructose-Gemischen mittels fixierter Glucoseoxidase und Katalase. Starch/Stärke 31: 348-353

Havewala N B, Weetall H H (1973) Foraminous containers in a stirrer shaft. US Pat no 3.767.535

Hofstee B H J (1973) Immobilization of enzymes through non-covalent binding to substituted agaroses. Biochem Biophys Res Commun 53: 1137-1144

Hossain M M, Do D D (1985) Fundamental studies of glucose oxidase immobilization on controlled pore glass. Biotechnol Bioeng 27: 842-851

Ibrahim M, Hubert P, Dellacherie E, Magdalou J, Muller J, Siest G (1985) Covalent attachment of epoxide hydrolase to dextran. Enz Microb Technol 7: 66-72

Imai K, Shiomi T, Uchida K, Miya M (1986) Immobilization of enzymes onto poly(ethylene-vinyl alcohol) membrane. Biotechnol Bioeng 28: 198-203

International Union of Biochemistry (ed) (1984) Enzyme nomenclature. Academic Press, London, 646 pp

Itoh N, Cheng L Y, Izumi Y, Yamada H (1987) Immobilized bromoperoxidase of Corallina pilulifera as a multifunctional biocatalyst. J Biotechnol 5: 29-38

Jennissen H P (1986) Protein binding to two-dimensional hydrophobic binding-site lattices: sorption kinetics of phosphorylase b on immobilized butyl residues. J Colloid Interface Sc 111: 570-586

Jack T R, Zajic J E (1977) The enzymatic conversion of L-histidine to urocanic acid by whole cells of Micrococcus luteus immobilized on carbodiimide activated carbxymethylcellulose. Biotechnol Bioeng 19: 631

Jirku V, Turkova J, Krumphanzl V (1980) Immobilization of yeast with retentin of cell division and extracellular production of macromolecules. Biotechnol Lett 2: 509-513

Karube I, Aizawa K, Ikeda S, Suzuki S (1979) Carbon dioxide fixation by immobilized chloroplasts. Biotechnol Bioeng 21: 253-260

Karube I, Kawarai M, Matsuoka H, Suzuki S (1985) Production of L-glutamate by immobilized protoplasts. Appl Microbiol Biotechnol 21: 270-272

Kato T, Horikoshi K (1984) Immobilized cyclomaltodextrin glucanotransferase of an alkalophilic Bacillus sp no 38-2. Biotechnol Bioeng 26: 595-598

Kaul R, D'Souza S F, Nadkarni G B (1984) Hydrolysis of milk lactose by immobilized ß-galactosidase-hen egg white powder. Biotechnol Bioeng 26: 901-904

Khan S S, Siddiqi A M (1985) Studies on chemically aggregated pepsin using glutaraldehyde. Biotechnol Bioeng 27: 415-419

Kobayashi T, Ohmiya K, Shimizu S (1975) Immobilization of ß-galactosidase by polyacrylamid gel. In: Weetall H H , Suzuki H (eds) Immobilized enzyme technology. Plenum Press, New York; pp 169-197

Koga J, Yamaguchi K, Gondo S (1984) Immobilization of alkaline phosphatase on activated alumina particles. Biotechnol Bioeng 26: 100-103

Krakowiak W, Jach M, Korona J, Sugier H (1984) Immobilization of glucoamylase on activated aluminiumoxide. Starch/Stärke 36: 396-398

Kubo I, Karube I, Suzuki S, Nambu Y, Endo T, Okawara M (1986) Substituted poly-methylglutamate membrane for immobilization of urease. J Biotechnol **4**: 27-33

Kühn W, Kirstein D, Mohr P (1980) Darstellung und Eigenschaften trägerfixierter Glukoseoxydase. Acta Biol Med Germ **39**: 1121-1128

Madry N, Zocher R, Grodzki K, Kleinkauf H (1984) Selective synthesis of depsi-peptides by the immobilized multienzyme enniatin synthetase. Appl Microbiol Biotechnol **20**: 83-86

Makryaleas K, Scheper T, Schügerl K, Kula M-R (1985) Enzymkatalysierte Dar-stellung von L-Aminosäure mit kontinuierlicher Coenzym-Regenerierung mittels Flüssigmembran-Emulsionen. Chem Ing Tech **57**: 362-363

Marek M, Valentova O, Kas J (1984) Invertase immobilization via its carbohydrate moiety. Biotechnol Bioeng **26**: 1223-1226

Mazumder T K, Sonomoto K, Tanaka A, Fukui S (1985) Sequential conversion of cortexolone to prednisolone by immobilized mycelia of Curvularia lunata and immo-bilized cells of Arthrobacter simplex. Appl Microbiol Biotechnol **21**: 154-161

Messing R A, Oppermann R A (1979) Pore dimensions for accumulating biomass. I. Microbes that reproduce by fission or budding. Biotechnol Bioeng **21**: 49-58

Mitz M A (1956) New insoluble active derivative of an enzyme as a model for study of cellular metabolism. Science **123**: 1076-1077

Miyama H, Kobayashi T, Nosaka Y (1984) Immobilization of enzyme on nylon con-taining pendant quaternized amine groups. Biotechnol Bioeng **26**: 1390-1392

Miyawaki O, Wingard jr L B (1984) Electrochemical and enzymatic activity of flavin dinucleotide and glucose oxidase immobilized by adsorption on carbon. Biotechnol Bioeng **26**: 1364-1371

Monsan P, Combes D (1984) Application of immobilized invertase to continuous hydrolysis of concentrated sucrose solutions. Biotechnol Bioeng **26**: 347-351

Monsan P, Combes D, Alemzadeh I (1984) Invertase covalent grafting onto corn stover. Biotechnol Bioeng **26**: 658-664

Mori T, Sato T, Tosa T, Chibata I (1972) Studies on immobilized enzymes. X. Preparation and properties of aminoacylase entrapped into acrylamide gel-lattice. Enzymologia **43**: 213-226

Nakajima H, Sonomoto K, Usui N, Sato F, Yamada Y, Tanaka A, Fukui S (1985) Entrapment of Lavendula vera and production of pigments by entrapped cells. J Biotechnol **2**: 107-117

Navarro J M, Durand G (1977) Modification of yeast metabolism by immobilization onto porous glass. Eur J Appl Microbiol Biotechnol **4**: 243-254

Nelson J M, Griffin E G (1916) Adsorption of invertase. J Am Chem Soc **38**: 1109-1115

Ogino S (1970) Formation of fructose-rich polymer by water-insoluble dextran-sucrase and presence of a glycogen value-lowering factor. Agric Biol Chem **34**: 1268-1271

Okita W B, Bonham D B, Gainer J L (1985) Covalent coupling of microorganisms to a cellulosic support. Biotechnol Bioeng **27**: 632-637

Papageorgiou G C, Lagoyanni T (1986) Immobilization of photosynthetically active cyanobacteria in glutaraldehyde-crosslinked albumin matrix. Appl Microbiol Bio-technol **23**: 417-423

Pastorino A M, Dalzoppo D, Fontana A (1985) Properties of sepharose-bound ß-lactamase from Enterobacter cloacae. J Appl Biochem **7**: 93-97

Qureshi N, Tamhane D V (1985) Production of mead by immobilized whole cells of Saccharomyces cerevisiae. Appl Microbiol Biotechnol **21**: 280-281

Raghunath K, Rao K P, Joseph U T (1984) Preparation and characterization of urease immobilized onto collagen-poly(glycidyl methacrylate) graft copolymer **26**:

104-109

Richter G, Heinecker H (1979) Conversion of glucose into gluconic acid by means of immobilized glucose oxidase. Starch/Stärke 31: 418-422

Romanovskaya V A, Karpenko V I, Pantskhava E S, Greenberg T A, Malashenko Y R (1981) Catalytic properties of immobilized cells of methane-oxidizing and methanogenic bacteria. In: Moo-Young M (ed) Advances in Biotechnology, vol 3. Pergamon, Toronto; pp 367-372

Schöpp W, Grunow M (1986) Immobilization of yeast ADH by adsorption onto polyaminomethylstyrene. Appl Microbiol Biotechnol 24: 271-276

Shankar V, Kotwal S M, Rao S S Yeast cells entrapped in low-gelling temperature agarose for the continuous production of ethanol. Biotechnol Lett 7: 615-618

Shimizu S, Morioka H, Tani Y, Ogata K (1975) Synthesis of coenzyme A by immobilized microbial cells. J Ferment Technol 53: 77-83

Suzuki H, Ozawa Y, Maeda H (1966) Studies on the water-insoluble enzyme. Hydrolysis of sucrose by insoluble yeast invertase. Agr Biol Chem 30: 807-812

Talsky G, Gianitsopoulos G (1984) Intermolecular crosslinking of enzymes. In: 3rd Eur Congr Biotechnol, vol 1. Verlag Chemie, Weinheim; pp 299-305

Tanaka A, Yasuhara S, Gelff G, Osumi M, Fukui S (1978) Immobilization of yeast microbodies and the properties of immobilized microbody enzymes. Eur J Appl Microbiol Biotechnol 5: 17-27

Tanaka A, Hagi N, Gellf G, Fukui S (1980) Immobilization of biocatalysts by prepolymer methods. Adenylate kinase activity of immobilized yeast mitochondria. Agric Biol Chem 44: 2399-2405

Tosa T, Mori T, Fuse N, Chibata I (1967) Studies on continuous enzyme reactions. I. Screening of carriers for preparation of water-insoluble aminoacylase. Enzymologia 31: 214-224

Tosa T, Mori T, Chibata I (1969) Studies on continuous enzyme reactions. IV. Enzymatic properties of the DEAE-Sephadex-aminoacylase complex. Agric Biol Chem 33: 1053-1059

Tsai Y-L, Schlasner S M, Tuovinen O H (1986) Inhibitor evaluation with immobilized Nitrobacter agilis cells. Appl Environm Microbiol 52: 1231-1235

Tsuchida T, Yoda K (1981) Immobilization of D-glucose oxidase onto a hydrogen peroxide permselective membrane and application for an enzyme electrode. Enzyme Microbial Technol 3: 326-335

Umemura I, Takamatsu S, Sato T, Tosa T, Chibata I (1984) Improvement of production of L-aspartic acid using immobilized microbial cells. Appl Microbiol Biotechnol 20: 291-295

Vilanova E, Manjon A, Iborra J L (1984) Tyrosine hydroxylase activity of immobilized tyrosinase on Enzacryl-AA and CPC-AA supports: stabilization and properties. Biotechnol Bioeng 26: 1306-1312

Vogel H J, Brodelius P (1984) An in vivo ^{31}P NMR comparison of freely suspended and immobilized Catharanthus roseus plant cells. J Biotechnol 1: 159-170

Weetall H H, Mason R D (1973) Studies on immobilized papain. Biotechnol Bioeng 15: 455-466

Wiegel J, Dykstra M (1984) Clostridium thermocellum: adhesion and sporulation while adhered to cellulose and hemicellulose. Appl Microbiol Biotechnol 20: 59-65

Workman W E, Day D F (1984) Enzymatic hydrolysis of inulin to fructose by glutaraldehyde fixed yeast cells. Biotechnol Bioeng 26: 905-910

Younes G, Breton A M, Guespinn-Michel J (1987) Production of extracellular native and foreign proteins by immobilized growing cells of Myxococcus xanthus. Appl Microbiol Biotechnol 25: 507-512

Further Literature to the Main Chapters

General Fundamentals

Barker S A (1980) Immobilized enzymes. In: Rose A H (ed) Microbial enzymes and bioconversions. Academic Press, London, pp 331-367

Bucke C (1983) Immobilized cells. Phil Trans R Soc Lond **B300**: 369-389

Chibata I, Tosa T, Sato T (1986) Immobilized cells and enzymes. J Mol Catalysis **37**: 1-24

Creighton T E (1986) Proteins - structures and molecular properties. Freeman, Oxford, 515 pp

Dellweg H (1984) Industrielle Erzeugung von primären Fermentationsprodukten in der Bundesrepublik Deutschland und in der Europäischen Gemeinschaft. Forum Mikrobiol (Sonderh Biotechnol) **7**: 4-11

Fersht A (1984) Enzyme structure and mechanism. Freeman, Oxford, 475 p

Godfrey T, Reichelt J (1983) Industrial enzymology: the applications of enzymes in industry. Macmillan Ltd., London, 600 pp

Hacking A J (1986) Economic aspects of biotechnology. Cambridge University Press, Cambridge, 306 pp

Hepner L, Male C (1983) Industrial enzymes - present status and opportunities. In: Lafferty R M (ed) Enzyme technology. Springer, Berlin Heidelberg New York, pp 7-8

Klibanov A M (1983) Immobilized enzymes and cells as practical biocatalysts. Science **219**: 722-727

Lasch J (1987) Enzymkinetik. Springer, Berlin Heidelberg New York, 148 pp

Rose A H (1980) History and scientific basis of commercial exploitation of microbial enzymes of bioconversions. In: Rose A H (ed) Microbial enzymes and bioconversions. Academic Press, London, pp 1-47

World Health Organization (1984) Health impact of biotechnology. Swiss Biotech **2** (5): 7-32

Zech R, Domagk (1986) Enzyme. Verlag Chemie, Weinheim, 184 pp

Methods of Immobilization

Beddows C G, Gil M H, Guthrie J T (1986) Immobilization of BSA, enzymes and cells of Bacillus stearothermophilus onto cellulose, polygalacturonic acid and starch based craft copolymers containing maleic anhydride. Biotechnol Bioeng **28**: 51-57

Cabral J M S, Novais J M, Kennedy J F (1986) Immobilization studies of whole microbial cells on transition metal activated inorganic supports. Appl Microbiol Biotechnol **23**: 157-162

Hulst A C, Tramper J, Riet K van't, Westerbeek J M M (1985) A new technique for the production of immobilized biocatalyst in large qantities. Biotechnol Bioeng **27**: 870-876

Klein J, Wagner F (1983) Methods for the immobilization of microbial cells. In: Chibata I, Wingard L B (eds) Applied biochemistry and bioengineering, vol **4**, Academic Press, New York, pp 11-51

Ito H, Shimizu A, Ichikizaki I (1986) Preparation and properties of stable water-insoluble derivatives of glutathione s-aryltransferase. Biotechnol Bioeng 28: 97-100

Kucera J (1986) The polymeric p- and o-quinones as the reactive supports for enzymes immobilization. Biotechnol Bioeng 28: 110-111

Kumakura M, Kaetsu I, Nisizawa K (1984) Cellulase production from immobilized growing cell composites prepared by radiation polymerization. Biotechnol Bioeng 26: 17-21

Marty J-L (1985) Application of response surface methodology to optimization of glutaraldehyde activation of a support for enzyme immobilization. Appl Microbiol Biotechnol 22: 88-91

Mattiasson B (1982) Immobilization methods. In: Mattiasson B (ed) Immobilized cells and organelles, vol 1. CRC Press, Boca Raton, pp 3-25

Mayer L D, Bally M B, Hope M J, Cullis P R (1986) Techniques for encapsulating bioactive agents into liposomes. Chem Phys Lipids 40: 333-345

Okita W B, Bonham D B, Gainer J L (1985) Covalent coupling of microorganisms to a cellulosic support. Biotechnol Bioeng 27: 632-637

Papisov M I, Maksimenko A V, Torchilin V P (1985) Optimization of reaction conditions during enzyme immobilization on soluble carboxyl-containing carriers. Enz Microb Technol 7: 11-16

Rouxhet P G, Haecht J L van, Didelez J, Gerard P, Briquet M (1981) Immobilization of yeast cells by entrapment and adhesion using silicious materials. Enzyme Microbial Technol 3: 49-54

Sikyta B (1986) Methods of cell immobilization. Microbiological Sci 3: 16-17

Tanaka H, Kurosawa H, Kokufuta E, Veliky I A (1984) Preparation of immobilized glucoamylase using Ca-alginate gel coated with partially quaternized poly(ethyleneimine). Biotechnol Bioeng 26: 1393-1394

Tsuge H, Okada T (1984) Immobilization of yeast pyridoxaminephosphate oxidase to halogenoacetyl polysaccharides. Biotechnol Bioeng 26: 412-418

Vorlop K D, Klein J (1983) New developments in the field of cell immobilization - formation of biocatalysts by ionotropic gelation. In: Lafferty R M (ed) Enzyme technology. Springer, Berlin Heidelberg New York, pp 219-235

Wang H Y, Lee S S, Takach Y, Cawthon L (1982) Maximizing microbial cell loading in immobilized cell systems. Biotechnol Bioeng Symp 12: 139-146

Wongkhalaung C, Kashiwagi Y, Magae Y, Ohta T, Sasaki T (1985) Cellulase immobilized on a soluble polymer. Appl Microbiol Biotechnol 21: 37-41

Characteristics of Immobilized Biocatalysts

Bourdillon C, Hervagault C, Thomas D (1985) Increase in operational stability of immobilized glucose oxidase by the use of an artificial cosubstrate. Biotechnol Bioeng 27: 1619-1622

Buchholz K (1982) Reaction engineering parameters of immobilized biocatalysts. In: Fiechter A (ed) Advances in biochemical engineering, vol 24. Springer, Berlin Heidelberg New York, pp 39-71

Buchholz K (1983) Parameters involved in heterogeneous biocatalysis. In: Lafferty R M (ed) Enzyme technology. Springer, Berlin Heidelberg New York, pp 9-21

Dagys R J, Pauliukonis A B, Kazlauskas D A (1984) New method for the determination of kinetic constants for two-stage deactivation of biocatalysts. Biotechnol Bioeng 26: 620-622

Davis M E, Watson L T (1985) Analysis of a diffusion-limited hollow fiber reactor for the measurement of effective substrate diffusivities. Biotechnol Bioeng **27**: 182-186

Do D D (1985) Determination of K^m and D^e from a one-shot experiment of enzyme immobilization. Biotechnol Bioeng **27**: 882-886

Hannoun B J M, Stephanopoulos G (1986) Diffusion coefficients of glucose and ethanol in cell-free and cell-occupied calcium alginate membranes. Biotechnol Bioeng **28**: 829-835

Hossain M M, Do D D (1985) Modeling of enzyme immobilization in porous membranes. Biotechnol Bioeng **27**: 1126-1135

Juang H-D, Weng H-S (1984) Performance of biocatalysts with nonuniformly distributed immobilized enzymes. Biotechnol Bioeng **26**: 623-626

Konecny J (1983) Kinetics and thermodynamics of reactions catalyzed by penicillin acylase-type enzymes. In: Lafferty R M (ed) Enzyme technology. Springer, Berlin Heidelberg New York, pp 309-314

Lenders J-P, Crichton R R (1984) Thermal stabilization of amylolytic enzymes by covalent coupling to soluble polysaccharides. Biotechnol Bioeng **26**: 1343-1351

Radovich J M (1985) Mass transfer effects in fermentations using immobilized whole cells. Enzyme Microbial Technol **7**: 2-10

Tanaka H, Matsumura M, Veliky I A (1984) Diffusion characteristics of substrates in Ca-alginate gel beads. Biotechnol Bioeng **26**: 53-58

Ulbrich R, Schellenberger A, Damerau W (1986) Studies on the thermal inactivation of immobilized enzymes. Biotechnol Bioeng **28**: 511-522

Vallat I, Monsan P, Riba J P (1985) Influence of glucose on the kinetics of maltodextrin hydrolysis using free and immobilized glucoamylase. Biotechnol Bioeng **27**: 1274-1275

Reactors for Immobilized Biocatalysts

Adler I, Fiechter A (1983) Charakterisierung von Bioreaktoren mit biologischen Testsystemen. Swiss Biotech **1**: 17-24

Andrews G F, Przezdziecki J (1986) Design of fluidized-bed fermentors. Biotechnol Bioeng **28**: 802-810

Ching C B, Ho Y Y (1984) Flow dynamics of immobilized enzyme reactors. Appl Microbiol Biotechnol **20**: 303-309

Chotani G K, Constantinides A (1984) Immobilized cell cross-flow reactor. Biotechnol Bioeng **26**: 217-220

Dale M C, Okos M R, Wankat P C (1985) An immobilized cell reactor with simultaneous product separation. II. Experimental reactor performance. Biotechnol Bioeng **27**: 943-952

Flaschel E, Raetz E, Renken A (1983) Development of a tubular recycle membrane reactor for continuous operation with soluble enzymes. In: Lafferty R M (ed) Enzyme technology. Springer, Berlin Heidelberg New York, pp 285-295

Goldstein L, Levy M (1983) Kinetics of multilayer immobilized enzyme-filter reactors: behavior of urease-filter reactors in different buffers. Biotechnol Bioeng **25**: 1485-1499

Guiot S R, Berg L van den (1985) Performance of an upflow anaerobic reactor combining a sludge blanket and a filter treating sugar waste. Biotechnol Bioeng **27**: 800-806

Hermanowicz S W, Ganczarczyk J J (1985) Mathematical modelling of biological packed and fluidized bed reactors. In: Joergensen S E, Gromiec M J (eds) Mathematical models in biological waste water treatment. Elsevier, Amsterdam, pp 473-524

Kloosterman J, Lilly M D (1985) An airlift loop reactor for the transformation of steroids by immobilized cells. Biotechnol Lett **7:** 25-30

Park T H, Kim I H (1985) Hollow-fibre fermenter using ultrafiltration. Appl Microbiol Biotechnol **22:** 190-194

Park Y, Davies M E, Wallis D A (1985) Analysis of a continuous aerobic fixed-film bioreactor. II. Dynamic behavior. Biotechnol Bioeng **26:** 468-476

Patwardhan V S, Karanth N, G (1982) Film diffusional influences on the kinetic parameters in packed-bed immobilized enzyme reactors. Biotechnol Bioeng **26:** 763-780

Shiotani T, Yamane T (1981) A horizontal packed-bed bioreactor to reduce CO_2 gas holdup in the continuous production of ethanol by immobilized yeast cells. Eur J Appl Microbiol Biotechnol **13:** 96-101

Wandrey C (1984) Bioreaktoren für den Einsatz von Enzymen. Forum Mikrobiol (Sonderh Biotechnol) **7:** 33-39

Yamane T, Shimizu S (1982) The minimum-sized ideal reactor for continuous alcohol fermentation using immobilized microorganisms. Biotechnol Bioeng **24:** 2731-2737

Zlokarnik M (1981) Verfahrenstechnische Grundlagen der Reaktorgestaltung. Acta Biotechnol **1:** 311-325

Industrial Applications

Boer R de, Romijn D J, Straatsma J (1982) Hydrolysed whey syrups made with immobilized ß-galactosidase in a fluidized-bed reactor. Neth Milk Dairy J **36:** 317-331

Borglum G B, Marshall J J (1984) The potential of immobilized biocatalysts for production of industrial chemicals. Appl Biochem Biotechnol **9:** 117-130

Cheetham P S J (1980) Developments in the immobilization of microbial cells and their application. In: Wiseman A (ed) Topics in enzyme and fermentation biotechnology, vol 4. John Wiley, New York, pp 189-238

Chibata I, Tosa T (1981) Use of immobilized cells. Ann Rev Biophys Bioeng **10:** 197-216

Chibata I, Tosa T, Sato T (1983) Immobilized cells in preparation of fine chemicals. Adv Biotechnol Processes **1:** 203-222

Hartmeier W (1985) Anwendung immobilisierter Biokatalysatoren. Chem Ind **37:** 321-323

Kennedy J F, Cabral J M S (1983) Immobilized living cells and their application. In: Chibata I, Wingard L B (eds) Applied biochemistry and bioengineering, vol 4. Academic Press, New York, pp 189-280

Leuchtenberger W, Karrenbauer M, Plöcker U (1984) Herstellung von L-Aminosäuren und entsprechenden a-Hydroxysäuren in einem Enzym-Membran-Reaktor. Forum Mikrobiol (Sonderh Biotechnol) **7:** 40-45

Linko P, Linko Y-Y (1983) Applications of immobilized microbial cells. In: Chibata I, Wingard L B (eds) Applied biochemistry and bioengineering, vol 4. Academic Press, New York, pp 53-151

Linko P, Linko Y-Y (1984) Industrial applications of immobilized cells. Crit Rev Biotechnol **1:** 289-338

Luong J H T, Tseng M C (1984) Process and technoeconomics of ethanol production by immobilized cells. Appl Microbiol Biotechnol **19:** 207-216

Messing R A (1982) Immobilized microbes and a high rate continuous waste processor for the production of high BTU gas and the reduction of pollutants. Biotechnol Bioeng **24**: 1115-1123

Nagashima M, Azuma M, Noguchi S, Inuzuka K, Samejima H (1984) Continuous ethanol fermentation using immobilized yeast cells. Biotechnol Bioeng **26**: 992-997

Roberts J (1985) Mathematical models for the trickling filter process. In: Joergensen S E, Gromiec M J (eds) Mathematical models in biological waste water treatment. Elsevier, Amsterdam, pp 243-324

Sayed S, Zeeuw W de, Lettinga G (1984) Anaerobic treatment of slaughterhouse waste using a flocculant sludge UASB reactor. Agric Wastes **11**: 197-226

Vandamme E J (1983) Immobilized enzyme and cell technology to produce peptide antibiotics. In: Lafferty R M (ed) Enzyme technology. Springer, Berlin Heidelberg New York, pp 237-270

Viraraghavan T, Landine R C, Winchester E L, Wasson G P (1984) Activated biofilter process for wastewater treatment. Effluent Water Treat J **24**: 378-384

Analytical Applications

Al-Hitti I K, Moody G J, Thomas J D R (1984) Glucose oxidase membrane systems based on poly(vinylchloride) matrices for glucose determination with an iodide ion-selective electrode. Analyst **109**: 1205-1208

Blum L J, Coulet P R, Gautheron D C (1985) Collagen strip with immobilized luciferase for ATP bioluminescent determination. Biotechnol Bioeng **27**: 232-237

Bowers L D (1986) Applications of immobilized biocatalysts in chemical analysis. Anal Chem **58**: 513A-530A

Bradley C R, Rechnitz G A (1986) Comparison of oxalate oxidase electrodes for urinary oxalate determinations. Anal Lett **19**: 151-162

Bowers L D, Carr P W (1980) Immobilized enzymes in analytical chemistry. In: Fiechter A (ed) Advances in biochemical engineering, vol **5**. Springer, Berlin Heidelberg New York, pp 89-129

Clark M F, Adams A N (1977) Characteristics of the microplate method of enzyme-linked immunosorbent assay for the detection of plant viruses. J gen Virol **34**: 475-483

Clarke D J, Blake-Coleman B C, Calder M R, Carr R J G, Moody S C (1984) Sensors for bioreactor monitoring and control - a perspective. J Biotechnol **1**: 135-158

Danielsson B (1983) Use of enzyme thermistor as a flow analyzer in biotechnology. In: Lafferty R M (ed) Enzyme technology. Springer, Berlin Heidelberg New York, pp 195-206

Guilbault G G (1981) Applications of enzyme electrodes in analysis. Ann New York Acad Sci **369**: 285-294

Janson J-C (1984) Large-scale affinity purification - state of art and future prospects. Trends Biotechnol **2**: 31-38

Karube I, Matsunaga T, Suzuki S, Asano T, Itoh S (1984) Immobilized antibody-based flow type enzyme immunosensor for determination of human serum albumin. J Biotechnol **1**: 279-286

Kashiwabara K, Hobo T, Kobayashi E, Suzuki S (1985) Flow injection analysis for traces of zinc with immobilized carbonic anhydrase. Anal Chim Acta **178**: 209-215

Kingdon C F M (1985) Biosensor design: microbial loading capacity of acetylcellulose membranes. Appl Microbiol Biotechnol **21**: 176-179

Kricka L J, Thorpe H G (1986) Immobilized enzymes in analysis. Trends Biotechnol **4**: 253-258

Liu M-T, Ram B P, Hart L P, Pestka J J (1985) Indirect enzyme-linked immuno-sorbent assay for the mycotoxin zearalenone. Appl Environ Microbiol 50: 45-49

Makovos E B, Liu C C (1985) Measurements of lactate concentration using lactate oxidase and an electrochemical oxygen sensor. Biotechnol Bioeng 27: 167-170

Margineanu D G, Vais H, Ardeleau I (1985) Bioselective electrodes with immo-bilized bacteria. J Biotechnol 3: 1-9

Mattiasson B, Mandenius C F, Danielsson B, Harlander P (1983) Computer control of fermentations with biosensors. Ann New York Acad Sci 413: 193-196

Pacakova V, Stulik K, Brabcova D (1984) Use of the Clark oxygen sensor with immobilized enzymes for determinations in flow systems. Anal Chim Acta 159: 71-79

Raghavan K G, Devasagayam T P A, Ramakrishnan V (1986) Immobilized enzyme brushes for clinical analyses: urea determination. Anal Lett 19: 163-176

Renneberg R, Schubert F, Scheller F (1986) Coupled enzyme reactions for novel biosensors. Trends Biochem Sci 11: 216-220

Scheller F W, Schubert F, Renneberg R, Müller H G, Jaenchen M, Weise H (1985) Biosensors: trends and commercialization. Biosensors 1: 135-160

Schügerl K (1985) Sensor-Meßtechniken in der biotechnologischen Forschung und Industrie. Naturwissenschaften 72: 400-407

Thompson R Q, Crough S R (1984) Stopped-flow kinetic determination of glucose and lactate with immobilized enzymes. Anal Chim Acta 159: 337-342

Wieck H J, Heider G H, Yacynych A M (1984) Chemically modified reticulated vitreous carbon electrode with immobilized enzyme as a detector in flow-injection determination of glucose. Anal Chim Acta 158: 137-141

Medical Applications

Goosen M F A, O'Shea G M, Gharapetian H M, Chou S, Sun, A M (1985) Optimization of microencapsulation parameters: semipermeable microcapsules as a bioartificial pancreas. Biotechnol Bioeng 27: 146-150

Jarvis A P, Grdina T A (1983) Production of biologicals from microencapsulated living cells. Biotechniques 1: 22-27

Klein M D, Langer R (1986) Immobilized enzymes in clinical medicine: an emerging approach to new drug therapies. Trends Biotechnol 4: 179-186

Lamberti F V, Sefton M V (1983) Microencapsulation of erytrocytes in Eudragit-RL-coated calcium alginate. Biochim Biophys Acta 759: 81-91

Lim F (ed) (1984) Biomedical applications of microencapsulation. CRC-Press, Boca Raton, 168 pp

Sakai T, Katsuragi T, Tonomura K, Nishiyama T, Kawamura Y (1985) Implantable encapsulated cytosine deaminase having 5-fluorocytosine-deaminating activity. J Biotechnol 2: 13-21

Senatore F F, Bernath F R (1986) Urokinase bound to fibrocollagenous tubes: an in vitro kinetic study. Biotechnol Bioeng 28: 58-63

Senatore F F, Bernath F R (1986) Parameters affecting optimum activity and stability of urokinase bound fibrocollagenous tubes. Biotechnol Bioeng 28: 64-72

Applications in Base Research

Bickerstaff G F (1984) Application of immobilized enzymes to fundamental studies on enzyme structure and function. In: Wiseman A (ed) Topics in enzyme and fermentation biotechnology, vol 9. Wiley, New York, pp 162-201

Büchner K H, Zimmermann U (1982) Water relations of immobilized giant algal cells. Planta **154**: 318-325

Clark D S, Bailey J E (1984) Characterization of heterogeneous immobilized enzyme subpopulations using EPR spectroscopy. Biotechnol Bioeng **26**: 231-238

Lenders J-P, Germain P, Crichton R R (1985) Immobilization of a soluble chemically thermostabilized enzyme. Biotechnol Bioeng **27**: 572-578

Mercer D G, O'Driscoll K F (1981) Kinetic modeling of a multiple immobilized enzyme system. II. Application of the model. Biotechnol Bioeng **23**: 2465-2481

Schäfer H-J (1987) Photoaffinity labeling and photoaffinity crosslinking of enzymes. In: Eyzaguirre J (ed) Chemical modification of enzymes: active site studies. Ellis Horwood, Chichester, pp 45-62

Siegbahn N, Mosbach K (1985) Covalent immobilization of the multienzyme enniatin synthetase. Biotechnol Lett **7**: 297-302

Walters S N, D'Silva A P, Fassel V A (1982) Immobilized enzyme system for the conversion of benz(a)pyrene to fluorescent metabolites. Anal Chem **54**: 2571-2576

Special Developments and Trends

Andersson E, Johansson A-C, Hahn-Hägerdal B (1985) a-Amylase production in aqueous two-phase systems with Bacillus subtilis. Enzyme Microbial Technol **7**: 333-338

Brink L E S, Tramper J (1985) Optimzation of organic solvent in multiphase biocatalysis. Biotechnol Bioeng **27**: 1258-1269

Brodelius P, Mosbach K (1982) Immobilized plant cells. In: Perlman D (ed) Advances in applied microbiology vol 28. Academic Press, New York, pp 1-26

Brodelius P, Nilsson K (1983) Permeabilization of immobilized plant cells resulting in release of intracellularly stored products with preserved cell viability. Eur J Appl Microbiol Biotechnol **17**: 275-280

Brouers M, Hall D O (1986) Ammonia and hydrogen production by immobilized cyanobacteria. J Biotechnol **3**: 307-311

Deshpande A, D'Souza S F, Nadkarni G B (1987) Coimmobilization of D-amino acid oxidase and catalase by entrapment of Trigonopsis variabilis in radiation polymerised polyacrylamide beads. J Biosci **11**: 137-144

Felten P von, Zürrer H, Bachofen R (1985) Production of molecular hydrogen with immobilized cells of Rhodospirillum rubrum. Appl Microbiol Biotechnol **23**: 15-20

Hahn-Hägerdal B (1983) Co-immobilization involving cells, organelles, and enzymes. In: Mattiasson B (ed) Immobilized cells and organelles, vol 2. CRC-Press, Boca Raton, pp 79-94

Hartmeier W (1985) Immobilisierte Biokatalysatoren auf dem Weg zur zweiten Generation. Naturwissenschaften **72**: 310-314

Hartmeier W (1985) Immobilized biocatalysts - from simple to complex systems. Trends Biotechnol **3**: 149-153

Jeanfils J, Loudeche R (1986) Photoproduction of ammonia by immobilized heterocystic cyanobacteria - effect of nitrite and anaerobiosis. Biotechnol Lett **8**: 265-270

Ku K, Kuo M J, Delente J, Wilde B S, Feder J (1981) Development of a hollow-fibre system for large-scale culture of mammalian cells. Biotechnol Bioeng **23**: 79-95

Lilly M D, Woodley J M (1985) Biocatalytic reactions involving water-insoluble organic compounds. Stud Org Chem **22**: 179-192

Lydersen B K, Pugh G G, Paris M S, Sharma B P, Noll L A (1985) Ceramic matrix for large scale animal cell culture. Bio/Technol **3**: 63-67

Makryaleas K, Scheper T, Schügerl K, Kula M-R (1985) Enzymkatalysierte Darstellung von L-Aminosäure mit kontinuierlicher Coenzym-Regenerierung mittels Flüssigmembran-Emulsionen. Chem Ing Tech **57**: 362-363

Martinek K, Berezin I V, Khmelnitski Y L, Klyachko N L, Levashov A V (1987) Enzymes entrapped into reversed micelles of surfactants in organic solvents: Key trends in applied enzymalogy (biotechnology). Biocatalysis **1**: 9-15

Mattiasson B (1983) Applications of aqueous two-phase systems in biotechnology. Trends Biotechnol **1**: 16-20

Miyawaki O, Wingard L B jr, Brackin J S, Silver R S (1986) Formation of propylene oxide by Nocardia corallina immobilized in liquid paraffin. Biotechnol Bioeng **28**: 343-348

Musgrave S C, Kerby N W, Codd G A, Stewart W D P (1982) Sustained ammonia production by immobilized filaments of nitrogen-fixing cyanobacterium Anabena 27893. Biotechnol Lett **4**: 647-652

Nilsson K (1987) Methods for immobilizing animal cells. Trends Biotechnol **5**: 73-78

Nilsson K, Mosbach K (1984) Peptide synthesis in aqueous-organic solvent mixtures with a-chymotrypsin immobilized to tresyl chloride-activated agarose. Biotechnol Bioeng **26**: 1146-1154

Rosevear A, Lambe C A (1985) Immobilized plant cells. In: Fiechter A (ed) Advances in Biochemical Engineering vol **31**. Springer, Berlin Heidelberg New York, pp 37-58

Spier R E (1980) Recent developments in the large scale cultivation of animal cells in monolayers. In: Fiechter A (ed) Advances in Biochemical Engineering vol **14**. Springer, Berlin Heidelberg New York, pp 119-162

Thomasset B, Barbotin J-N, Thomas D (1984) The effects of high concentrations of salts on photosynthetic electron transport of immobilized thylakoids: functional stability. Appl Microbiol Biotechnol **19**: 387-392

Tschopp A, Cogoli A, Lewis M L, Morrison D R (1984) Bioprocessing in space: human cells attach to beads in microgravity. J Biotechnol **1**: 287-293

Verlaan P, Hulst A C, Tramper J, van't Riet K, Luyben K C A M (1984) Immobilization of plant cells and some aspects of the application in an airlift loop reactor. In: 3rd Eur Congr Biotechnol. Verlag Chemie, Weinheim, pp 151-154

Index